Advances in Data Analytics, AI, and Smart Systems

Series Editor
Robin Qiu, Division of Engineering & Info Sci
Pennsylvania State University
Malvern, PA, USA

This book series investigates the transformative influence of data analytics and artificial intelligence (AI) on modern smart systems. Driven by the convergence of big data, AI, the Internet of Things (IoT), cloud computing, and Web 3.0, these technologies are revolutionizing industries and reshaping societal structures. The series critically examines how these innovations optimize processes, enhance decision-making, and drive cross-sector innovation.Beyond core themes like AI/IoT integration and interdisciplinary applications (e.g., healthcare, finance, smart cities), the series prioritizes emerging trends, including Generative AI, Artificial General Intelligence (AGI), blockchain, quantum computing, and the decentralized potential of Web 3.0. Crucially, it also highlights ethical and governance considerations, fostering research on responsible AI development, data security and privacy, and bias mitigation in algorithmic decision-making.

This series publishes monographs, contributed volumes, and professional books that provide firsthand insights into the latest research, pioneering innovations, and new perspectives on classical fields. Its target audience includes students, researchers, and industry professionals.

Robin Qiu

Value Co-Creation in Web 3.0

Blockchain, Metaverse, and Privacy Preservation

Springer

Robin Qiu
Pennsylvania State University
University Park, PA, USA

ISSN 3091-2733 ISSN 3091-2741 (electronic)
Advances in Data Analytics, AI, and Smart Systems
ISBN 978-3-032-12910-9 ISBN 978-3-032-12911-6 (eBook)
https://doi.org/10.1007/978-3-032-12911-6

This Springer imprint is published by the registered company Springer Nature Switzerland AG
The registered company address is: Gewerbestrasse 11, 6330 Cham, Switzerland

Preface

Web 3.0, the future Internet, redefines the value of digital data assets, not as a commodity to be extracted, but as a highly rewarding outcome of genuine collaboration, collective co-governance, and continuous co-evolution of all individual participants worldwide. Blockchain ensures trust. The metaverse makes it experiential. Privacy preservation makes it ethical. Promisingly, the resulting smart *deepservice* makes it profoundly human. This is not merely the future of technology; it is the definitive future of smart living.

In the summer of 2024, I started writing this book during my leave of absence from Penn State. I spent over six months at the Zi Ka Wei library in Shanghai, reflecting on the profound digital transformations through the Internet the world has seen since the 1990s. While I was there, I watched Shanghai, in particular the downtown area of Zi Ka Wei evolved, becoming increasingly modern, digitalized, and AI-enabled. This local transformation solidified my belief that the rest of the world was on the same path.

Truly, the Internet has profoundly transformed human life, arguably more than any other technology in the past half-century. Although it was originally conceived as a simple research network, it has evolved into a global infrastructure permeating every aspect of our economies, societies, and daily lives. Yet, this evolution is far from over. As digital economies experience unprecedented growth, we are entering a new information era where the convergence of Web 3.0 and AI is set to fundamentally reshape how we create, share, and govern digital value.

This book explores the new information era: Web 3.0 and AI. Built on the foundations of decentralization, blockchain, privacy-preserving computation, massive and rich data, algorithms, and intelligent automation, Web 3.0 seeks to redefine our digital lives around the core principles of trust, autonomy, transparency, and equitable value co-creation. It challenges the limitations of the platform-centric Web 2.0 era, replacing centralized control with user sovereignty, data exploitation with data stewardship, and passive consumption with active participation. By incorporating the ever-growing AI into Web 3.0, we can develop personalized and context-aware digital services at the point of demand, largely focusing on meeting the needs of end-users.

However, technology alone does not guarantee a better digital future. As history has shown, the true value of technological innovation depends on how it is applied and who benefits over time. This is why this book frames Web 3.0 within the principles of

Service Science, an interdisciplinary approach to understanding and designing human-centric service systems that co-create value for all stakeholders. This vision articulates that smart service ecosystems must be not only human-centric but ultimately beneficial to human beings, considering equality, prosperity, and ethicality. In this light, the Web 3.0 ecosystem is far beyond a collection of technologies, computational sciences, and protocols, but a living, evolving service ecosystem where humans, organizations, intelligent machines, and AI agents collaborate for shared benefit, building a human-centered, brighter, and more sustainable future.

The ensuing chapters elaborate on the evolutionary journey from the Internet's origins to the nascent architecture of Web 3.0, with an emphasis on its transformative potential and inherent challenges. This exploration encompasses an examination of foundational technologies, specifically blockchain, smart contracts, and privacy-preserving computation. We define decentralized identity, decentralized finance, decentralized governance, and decentralized applications as the pillars of Web 3.0, which collectively serve as the core of the functional operating system that will support and enable the future of Web 3.0. Through two detailed case studies, we demonstrate how these elements can be effectively integrated into real-world domains, such as smart mobility and human-centric smart cities. Finally, we look ahead to how the convergence of AI, trusted data spaces, and the metaverse will usher in the next generation of smart *deepservice*: intelligent, personalized, context-aware, and ethically grounded services designed for a truly human-centered digital era.

This is not a book about hype; it is a book about possibilities—possibilities grounded in technical reality, informed by interdisciplinary thinking, and guided by the belief that technology must serve human needs above all else. Whether you are a researcher, policymaker, entrepreneur, or simply a curious reader, I invite you to explore how Web 3.0, when aligned with the principles of Service Science, can help co-create a digital future that is more open, fairer, and more deeply human.

University Park, PA, USA Robin Qiu
October 10, 2025

Acknowledgments

The author would like to acknowledge the help of all involved in the collation and review process of this book; without their support, the project could not have been satisfactorily completed. Special thanks also go to all the staff at Springer, Inc., whose contributions throughout the process—from the initial idea to final publication—have been invaluable. In particular, I am very grateful to **Jialin Yan**, my book editor, who continuously prodded me via e-mail to keep the project on schedule and provided various helps throughout my book-writing journey. I also thank the **anonymous reviewers** whose insights and criticisms significantly improved the quality of this book.

Finally, I want to thank my wife, Lihua Zhang, and children, Lawrence and Jason, for their love and support throughout this project. This book is dedicated to my family!

Robin G. Qiu

Disclaimer No product or service mentioned in this book is endorsed by its maker or provider, nor are any claims of the capabilities of the product or service discussed or mentioned. The names of products and organizations mentioned if any may be the trademarks or registered trademarks of their respective owners.

Contents

The Evolution of the Internet

1

Abstract

In this chapter, we chart the Internet's journey, beginning with its genesis in ARPANET and TCP/IP and moving through the establishment of the World Wide Web and early e-commerce. We chronicle the seismic shift brought by Web 2.0, defined by social networking and mobile computing, and arrive at the current landscape driven by the fusion of IoT, 5G/6G, and cloud computing, which power modern data and platform economies. However, the core of our discussion explores Web 3.0 (Web3), a decentralized, human-centered future. As a paradigm built on blockchain, privacy-preserving computation, AI, and the Metaverse, Web3 promises to fundamentally redefine the digital ecosystem by prioritizing user control, transparency, and equity. Crucially, we contend that the principles of Service Science offer the ideal framework to maximize ethical, data-driven value co-creation and innovation within this new era. This chapter establishes that Web3 represents a revolutionary change in value co-creation and exchange, not merely an incremental technological update, setting the stage for the rest of the book.

Keywords

Internet · World Wide Web (WWW) · Web 2.0 · Web 3.0 · Internet of Things (IoT) · Wireless networks · Cloud computing · Service Science

1.1 Introduction

Since the 1990s, we have witnessed rapid and significant improvements in product and service quality, along with substantially increased productivity in manufacturing and engineering, attributed mainly to efficient resource allocation worldwide and to unprecedented, innovative business models, driven largely by Internet

R. Qiu, *Value Co-Creation in Web 3.0*, Advances in Data Analytics, AI, and Smart Systems, https://doi.org/10.1007/978-3-032-12911-6_1

technology. The twenty-first century world has undoubtedly long been preparing for the coming of a full-blown digital economy, which is well understood to rely substantially on advanced sciences and technologies, particularly Internet technology, to enhance all aspects of future economic activities.

Notably, the significance of the development of Internet technology can hardly be described using a few words, given that people cannot imagine how they could live, work, and travel if there were no Internet today. Indeed, the Internet has transformed nearly everything in our daily lives, wherever we are. The following limited list highlights some of the key transformations that have occurred over the last several decades:

- Information access and sharing: Thanks to the Internet, a vast array of data, information, and knowledge—ranging from daily news and entertainment to archived documents, social media, and most recently published academic research and discoveries—has been made readily accessible and shareable anywhere, at any time.
- Communication and connectivity: Regardless of time and location, the Internet has enabled instant messaging, video conferencing, and social networking. As a result, we can now instantly communicate with our family, friends, colleagues, and business partners. In the professional world, the enabled efficient connectivity and fast communication have facilitated the growth of online services, virtual communities, international trade, and, in general, global collaboration across businesses, industries, and societies.
- Economic development: The Internet has been one of the major drivers of economic growth, as new businesses such as e-commerce, digital marketing, platform services, cloud computing, and telecommunications have been rapidly created, developed, and improved. These new businesses have substantially transformed various sectors, including governmental and public services, retail, healthcare, tourism, education, training, banking, and entertainment. There have been countless opportunities and invaluable benefits created for entrepreneurs and customers. Today, it is widely recognized that significantly improved capital efficiency and labor productivity have had an enormous and unprecedented impact on the global economy.
- Education and Research: The Internet has provided students, scholars, and researchers with easy access to online courses, digital publications, digital libraries, and scientific tools. Collaborative and cooperative research on a global scale has become increasingly productive and a trend. Over the past decade, the increasing accessibility and shareability of learning resources have vastly expanded educational opportunities, enabling them to reach remote and underrepresented communities worldwide through online education and distance learning.

Indeed, the Internet has played an essential role in our daily lives, shaping our culture, economy, and society in unprecedented ways. However, we have observed that the Internet has also presented a lot of challenges and risks, including political propaganda and conspiracy theories, privacy concerns, ethical issues, cyberbullying, misinformation, pornography, and online gaming and gambling addiction. It is necessary to address these challenges and risks effectively so that the Internet can be fully harnessed for the long-term benefit of humanity.

Currently, we understand data is primarily generated and derived from the Internet and from millions of applications, largely powered by Internet technologies. As data is not just a byproduct of modern life, it is a fundamental resource like coal and oil in the past industrial age. In fact, massive data has become the lifeblood of the current information and artificial intelligence era. How to leverage the ever-growing volume of data defines our collective future. However, to understand how the Internet will further evolve, it is necessary to appreciate its fascinating and challenging journey since its inception. Let us start by briefly reviewing how the Internet has been developed over the years.

The origin of the Internet has been mainly traced to the Advanced Research Projects Agency Network (ARPANET), a project started by the US Department of Defense in 1969. It was developed to enable data transmission among geographically dispersed computers, primarily for military and academic purposes related to resource-sharing. The first ARPANET consisted of only four computer hosts. In the 1970s, packet switching, and the Transmission Control Protocol and the Internet Protocol (TCP/IP) were developed as a result of the successful completion of the ARPANET project. They quickly became the fundamental networking technologies of the Internet (Leiner et al., 2009).

Technically, packet switching is a method of transmitting data by dividing it at the origin into small, independent packets that can travel along different network routes and be reassembled at the destination (Roberts, 1978). TCP/IP warrants that data transmission between computing devices in a heterogeneous network can be successfully implemented using packet switching (Cerf & Kahn, 1974). These protocols have since become the standard for network data transmission, laying the groundwork for the development of later Internet technology.

In the 1980s, the Domain Name System (DNS) was developed to map numeric IP addresses to human-readable domain names. DNSs made it easy and intuitive for users to navigate over interconnected networks (Mockapetris & Dunlap, 1988). Initially, the National Science Foundation (NSF) funded several supercomputer centers across the USA. To support collaborative research and information sharing across the US higher education and research communities, the NSF developed the NSF Network (NSFNET) in the late 1980s, which became a key backbone for the development of the Internet (Claffy et al., 1994).

To the public, the early Internet was centered on the World Wide Web, invented in 1989, which aimed to enable automated information sharing among interconnected computers across the higher education and research communities worldwide (Berners-Lee et al., 1992). Because the World Wide Web provided a user-friendly way to access information on the Internet, it quickly gained popularity among the public with the emergence of web browsers such as Mosaic (1993) and Netscape Navigator (1994) (Berghel, 1998).

I still vividly recall that in 1994, the Computer Integrated Manufacturing (CIM) research lab at Penn State, where I was a research assistant and pursuing my Ph.D., published our research information online through a series of static web pages. At that time, there were very few research-oriented websites globally. As a result, our online presence quickly garnered significant attention within the research community. My Ph.D. work culminated in the development of a fully automated CIM system capable of simultaneously producing five different product types by coordinating multiple numerically controlled machines, robots, a warehouse, and a conveyor system (Qiu, 1996). We also conducted live demonstrations over the Internet, enabling viewers from around the world to observe the production process in real time. The products manufactured by the fully automated CIM system were often given as souvenirs to visitors, many of whom came to the lab to learn from what was recognized as one of the leading CIM research facilities globally at the time.

The beginning of commercial use of the Internet started in the middle of the 1990s, which was pioneered by companies such as Yahoo's information sharing website launched in Jan. 1994 (Yahoo, 2025), Amazon's e-commerce platform launched in July 1994 (Amazon, 2025), eBay's auction service website launched in Sept. 1995 (eBay, 2025), and Google's effective and efficient PageRank-based search engine platform developed in Sept. 1998 (Brin & Page, 1998).

However, Wall Street had overinvested in the widespread adoption of the Internet for online shopping, communication, and news services, without considering the sustainability of their businesses. The stock market quickly rose and was characterized as "irrational exuberance" due to a surge in Internet-based companies. The dotcom bubble burst in 2000 (Ofek & Richardson, 2003). It certainly indicated that the world needed a better Internet to sustain dotcom businesses.

Instead of using static Web pages with vast amounts of information accessible to the public, the emergence of engaging people for better user experience and social networking on the Internet quickly marked the new era of the Internet—Web 2.0. Unlike the first generation of the Internet, which focused on information sharing and online sales, Web 2.0 platforms have capitalized on promoting user-generated content, interactivity, and social networking, thereby contributing to the platform economy while maintaining sustainability. As of 2025, the number of Internet users continued to grow, reaching approximately 5.5 billion, or roughly 68 percent of the global population (Statista, 2025a).

It is worth mentioning that MySpace, launched in 2003, was the first popular platform to provide a social networking service, reaching a global audience and revolutionizing the development of social networking technology, cultural understanding, and music distribution on the Internet (Boyd & Ellison, 2007). With the widespread adoption of broadband Internet and the rapid development of mobile computing, the Internet has grown rapidly as people around the world are increasingly engaged, anywhere, at any time. Notably, the Palm Pilot, BlackBerry, and iPhone, as personal digital assistant (PDA) products, have made an impressive mark in history since their inception.

In 1996, Palm OS, the first mobile operating system introduced for Palm Computing's Pilot PDA, was later adopted across a diverse range of mobile devices, including smartphones, smartwatches, gaming consoles, barcode scanners, and GPS units (Palm, 2025). In 1999, BlackBerry introduced push email and messaging capabilities, which quickly gained popularity among business users. Then in 2007, Apple introduced the iPhone, ushering in the modern smartphone era. Apple's iPhone was unveiled with its touchscreen interface and App Store, transforming mobile usage, catalyzing vast mobile app ecosystems, and redefining digital lifestyles (Firtman, 2013; Molla, 2017).

Today, we fully understand that platforms supporting diverse mobile app ecosystems, such as MySpace (2003), Facebook (2004), YouTube (2005), Twitter (2006), WhatsApp (2009), Instagram (2010), WeChat (2011), and TikTok (2017), are integral to people's daily lives as those social networking platforms have revolutionized the ways in which people, businesses, and organizations communicate, share data and information, and collaborate both at home and in the workplace (McDonald, 2022; TSM, 2025).

1.2 The Ever-Increasing Digitalized World

Because of the rapid evolution of digital technology, the world has become more digitized and interconnected than ever before over the past three decades. Technically, it can be attributed to the advances in computing, networking, the Internet, the Internet of Things (IoT), and mobile computing. The advances in these highly adopted digital technologies are widely recognized as the driving force propelling the accelerated global digital transformation. As mentioned earlier, mobile computing devices, especially smartphones and tablets, have fundamentally reshaped people's daily lives. We now routinely use mobile computing devices to shop, pay bills, manage travel, read the news, and communicate with loved ones from anywhere at any time.

It is worth noting that mobile computing is currently largely powered by iOS and Android, the two dominant mobile operating systems (Tracy, 2012). Apple's iOS is a proprietary mobile operating system designed exclusively for Apple's ecosystem of devices. As of 2025, Apple's iOS remains the core operating system powering the iPhone, iPad, and Apple Watch. It is well known that Apple's iOS enables unified, secure user experiences across these products. In contrast, Android is an open-source operating system first released by Google in 2008. It has become the most widely adopted operating system on mobile computing devices globally. Supported by a vast ecosystem of developers and apps, Android is poised to continue driving innovation and adapting to the latest advances in computing and data science.

By the first quarter of 2025, approximately 72% of smartphones were running on the Android Open Source Project (AOSP) and were operational within Google's ecosystem (Statista, 2025b). AOSP supports a large pool of vendor-customized utilities and software development kits (SDKs). As competition intensifies, there are also many forks of AOSP, including well-known ones such as Fire OS by Amazon, ColorOS by Oppo, OriginOS by Vivo, MagicUI by Honor, and Meta Horizon OS for VR/AR headsets (Android, 2025; Reveron, 2024).

In addition, the Android source code has been used to develop variants for a range of computing devices, including game consoles, digital cameras, portable players, wearables, TV top boxes, Tablets, mini-PCs, and a vast array of IoT devices. With the enablement of several million apps, it has supported over several billion monthly active users globally. As a result, it is currently the most widely installed operating system worldwide (Abdul Kadir et al., 2024; Android, 2025; Statista, 2025b).

Since 2000, three rapidly advancing computing and networking technologies have been powerful driving forces behind the accelerated digital transformation of businesses and societies worldwide. Those three computing and networking technologies are as follows:

- IoT (Internet of Things): making billions of devices equipped with network-based connectivity
- Wireless and 5G/6G technology: ensuring that not only physical devices but also people can communicate well without physical connections
- Cloud computing: expanding organizations' capabilities by enabling their users to access the needed data, information, and functions anywhere at any time.

Fundamentally, IoT is a network framework for enabling applications involving networked physical and virtual objects. By relying on various networks, including telecommunication networks, wireless communication, sensor networks, and the Internet, billions of objects, such as people, can be well connected. As a result, IoT enables data exchange and communication between objects, supporting intelligent identification, positioning, tracking, monitoring, and management. Today, smart homes, smart cities, wearable smart devices, and connected vehicles are quite common, integrating the Internet into more aspects of people's daily lives, leading to the emergence of a variety of new innovative services and applications we had never seen before, such as intelligent manufacturing, smart transportation, and smart healthcare (Wang et al., 2021).

Wireless technology focuses on transferring data and information over distances without physical wires, enabling greater flexibility and convenience for applications that would otherwise be difficult or impossible (Alam et al., 2023). Common types of wireless technology include Wi-Fi, Bluetooth, NFC (near fields communication), RFID (radio frequency identification), Zigbee & Z-Wave, and cellular networks. Since the 2010s, we have seen rapid advances in wireless communication speed, range, security, and energy efficiency. This is particularly true with the introduction of 5G wireless cellular networks in the early 2020s, which has driven their adoption across various industries, businesses, societies, and people's daily lives.

Currently, people around the world are enjoying the widespread use of social media networks such as Facebook, Instagram, Snapchat, WhatsApp, WeChat, and TikTok, as well as streaming entertainment services like Netflix and Spotify. The ongoing evolution from 5G to 6G paves the way for a hyper-connected future, unlocking the full potential of emerging technologies like augmented reality (AR), extended reality (XR), and the metaverse. People will soon start to embrace significantly enhanced immersive, intelligent, and cross-reality experiences at an unprecedented scale.

Unlike traditional on-premises computing, cloud computing offers businesses and end-users significant advantages in infrastructure, scalability, cost-effectiveness, and accessibility (Kiswani et al., 2021). With advances in computing and networking security and dependency, the 2010s witnessed the rapid growth of cloud computing, enabling enterprise data and applications to be reliably controlled and accessed over the Internet. The unceasingly improved flexibility and cost-effectiveness justify the widespread adoption of cloud computing across all applications.

As of the first half of 2025, cloud computing is pervasive in organizations. It has largely moved beyond being just an emerging technology, becoming the foundational infrastructure for most organizations worldwide. The well-adopted cloud computing platforms can be briefly listed as follows:

- Amazon Web Services (AWS) is currently the largest cloud computing service provider globally. AWS delivers a comprehensive suite of cloud capabilities, focusing on enabling services such as computation, storage, databases, and e-commerce platform utilities. Amazon Web Services is undoubtedly recognized for its high scalability and flexibility because of its global presence and a wider selection of computing and data centers.
- Microsoft Azure, as a cloud computing platform, focuses on providing a broad set of office services, particularly Microsoft Office products for computing, analysis, storage, and networking. Thus, Azure naturally becomes the best choice when a business has the central need for intensive integration with Microsoft's Office products, such as Office 365, SharePoint, Windows Servers, and Microsoft Copilot.
- Google Cloud, a cloud computing service, is renowned for its robust capabilities in artificial intelligence (AI) (e.g., Gemini), machine learning, data processing, and analysis. Google Cloud offers services like computing engines, data storage, machine learning, and data processing and analytics. As Google has been providing extensive analytics and integration tools and services, Google Cloud can be seamlessly integrated with not only Google's suite of online tools and services but also third-party applications to support digital transformation across business and society, with a focus on maximizing value for all stakeholders.
- IBM has been a provider of enterprise-grade computing solutions for over 100 years. IBM has made significant investments in areas such as virtualization, containers, AI, and data analytics. Hence, IBM Cloud has become known for its strong on-premises integration and flexible hybrid cloud solutions. Organizations choosing IBM Cloud aim to deploy enterprise-level, customer-centric infrastructures that meet their long-term, reliable, controllable, and scalable IT needs.
- Two well-known Chinese IT players include Alibaba and Tencent. Alibaba Cloud delivers a comprehensive suite of cloud services, including elastic computing, databases, storage, security, and AI, serving customers globally. Likewise, Tencent Cloud offers a range of services, including cloud servers, databases,

storage, and AI solutions. However, Tencent Cloud is well-known for its substantial presence in markets such as gaming, social media, and video streaming, particularly in China.

Indeed, IoT enables billions of devices and users to connect to the network. Wireless and 5G/6G technologies ensure seamless connection for these devices and users, even in physically inaccessible locations. Cloud computing services enable organizations to rapidly scale their digital applications. Moreover, the integration of IoTs, 5G/6G networks, and cloud computing technologies has significantly enhanced the architecture and capabilities of the Internet. Collectively, they enable low-latency, high-bandwidth, ubiquitous, pervasive, and scalable solutions that have been redefining people's social dynamics and business operational models across both personal and professional domains.

Notably, the digital transformation of work and life has been accelerating at an unceasing pace. According to Arnold et al. (2023), the world now generates as much data and information in 2 days as it did from the dawn of civilization until 2003. The enormous amount of data is widely considered the lifeblood of the modern digital economy. Thus, appropriately leveraging the massive amount of data becomes essential for businesses and societies to unlock operational insights and drive innovation for a prosperous future.

Today, this surge in data generation is accelerating the development of cutting-edge technologies and services, from AI and machine learning to data-driven decision-making and personalized experiences. The global excitement around the emerging Web 3.0 reveals the growing recognition of data's importance (GVR, 2025). This innovative Web 3.0 framework promotes and fosters a decentralized cyber ecosystem in which individuals, rather than organizations, are encouraged and empowered to control their digital assets and personal information. In other words, by leveraging technologies such as blockchain and privacy-preserving computation, Web 3.0 aims to create a more equitable and secure digital world than the centralized platforms of Web 2.0. It becomes crystal clear that as the Internet continues to evolve with advances in AI, 5G/6G, and IoT, Web 3.0 is poised to usher in a revolutionary shift toward developing a more human-centered digital future.

1.3 The Coming of the Web 3.0 Era

As the Internet continues to evolve, we are approaching a new frontier: Web 3.0. Driven by technological innovation and individual and societal needs for privacy and data security, the decentralized, human-centered Web 3.0 paradigm marks a significant departure from its predecessor, Web 2.0. According to Ghosh and Gorai (2024), Web 3.0 aims to empower individual users and promote data-driven decision-making in businesses and industries like never before. A fundamental change in Web 3.0 is the ability for individuals to own and monetize their data, while safeguarding their digital identities and personal information. By focusing on individuals' empowerment, this fundamental shift will certainly foster innovation and unlock new economic opportunities. In addition, by eliminating intermediaries and streamlining operational processes, Web 3.0-empowered applications will help reduce costs and increase transparency, leading to greater efficiency, security, and trust.

Although Web 3.0 is still emerging, its potential to fundamentally change businesses, industries, and societies becomes crystal clear (Ray, 2023; Wan et al., 2024). As we approach this new digital landscape, three key technological trends are likely to define its trajectory:

- A new digital paradigm focusing on decentralization: The digital revolution continues its march toward decentralization. Technologies such as blockchain, smart contracts, and decentralized applications (DApps) will help remove centralized power from large corporations in Web 2.0, empowering individuals and communities to foster independence and unprecedented potential. In other words, Web 3.0 will give users more control over their data and digital identities, encouraging users' participation and promoting new business opportunities and growth models.
- Intelligent evolution through AI and machine learning: The ongoing evolution of these technologies is reshaping how we engage with the digital world, enhancing personalization, efficiency, and intelligence across digital experiences. For example, from personalized content recommendations to sophisticated autonomous systems, these technologies will make our digital experiences more seamless, intuitive, and personal. As AI becomes more integrated into our daily lives, we can expect a future where intelligent assistants, smart homes, and self-driving vehicles are commonplace.
- Metaverse as a new virtual frontier: The concept of the metaverse, a collective, immersive virtual space, is quickly gaining momentum. Companies are investing heavily to create online environments that blur the lines between our physical and digital lives. As these virtual worlds mature, they will present new opportunities for social interaction, entertainment, and business, thereby creating innovative ways for people to connect and collaborate.

Ultimately, the success of Web 3.0 will depend on its ability to serve human needs. By prioritizing user experience, accessibility, and ethical considerations, we can use this technology to build a more equitable and fulfilling digital future for everyone.

We must clearly understand that Web 3.0 emerges not just as a technological upgrade, but as a transformative shift in how data, services, and value are structured, created, and exchanged. While its complete form is still materializing, the convergence of decentralization, intelligent automation, and immersive virtual environments is already redefining communication, commerce, and social interaction.

In summary, as the digital landscape continues to evolve, the Web 3.0 era is approaching. At the heart of this transformation lies a simple but powerful principle: technology must ultimately serve human needs. With blockchain enabling decentralized trust, AI and machine learning enhancing intelligent personalization, and the metaverse expanding experiential possibilities, Web 3.0 offers a new architecture for digital life. Yet beyond infrastructure and tools, the true potential of Web 3.0 lies in its capacity to empower value co-creation among individuals, businesses, and organizations. This next section highlights the theme and organization of this book, showing how Web 3.0, aligned with principles from Service Science, lays the foundation for deep, data-driven, and user-centered value co-creation in the digital economy (Qiu, 2024b).

1.4 Value Co-creation in Web 3.0

Web 3.0 technology empowers individuals with data self-control, revolutionizing data ownership and utilization. By collecting and harnessing data responsibly and ethically, we can drive significant social and economic progress (Wan et al., 2024). This is particularly crucial in the AI era, where trustworthy, reliable, and capable AI solutions are essential for addressing human needs (Qiu, 2024a). Hereafter, Web 3.0 will be referred to simply as Web3, and the terms will be used interchangeably.

Just as past technological revolutions, the rise of Web3, the successor to the traditional Internet (i.e., Web 2.0), presents both great potential and significant risks. To ensure its sustainable growth, fostering cross-border dialogue will be essential. Qiu (2024b) notes that Web3 and Service Science share a common goal: promoting and delivering value to all stakeholders.

This book, therefore, aims to explore how principles of Service Science can be integrated into the Web3 ecosystem. We will propose a corresponding framework that leverages the advancements in blockchain, the metaverse, and privacy-preserving technologies to maximize value co-creation in the Web3 era, ultimately serving human needs.

In the remaining chapters, this book will be organized as follows:

In Chap. 2, we will review the fundamental technologies that underpin the foundational pillars of the Web3 ecosystem. These include decentralization, where data is distributed across a network rather than stored on a single server, thereby enhancing security and resilience. We will also cover immutability, a core blockchain property that ensures data, once recorded, cannot be altered, thereby guaranteeing transparency and auditability. The public visibility of transactions on the blockchain fosters trust and accountability, though privacy levels can be adjusted through specific architectures and privacy-enhancing technologies. Finally, we will examine the high level of security provided by advanced cryptographic methods such as hashing, digital signatures, and privacy-preserving computations, which make it exceedingly difficult to compromise the data's integrity.

In Chap. 3, we will discuss how the Web3 ecosystem thrives on the seamless coordination of its foundational pillars: Decentralized Autonomous Organizations (DAOs), Decentralized Identifiers (DIDs), Decentralized Finance (DeFi), and Decentralized Applications (DApps). DAOs revolutionize governance through decentralized, transparent decision-making, while DIDs redefine digital identity with a secure, self-sovereign, and verifiable framework that reduces reliance on centralized authorities. DeFi transforms traditional financial systems by promoting equitable value distribution and transparency through decentralized ledgers. DApps embody Web3 principles by shifting users from passive consumers to active contributors in governance, development, and value co-creation, including monetization. Together, these components function as the pillars of Web3, ensuring a more decentralized, efficient, and user-driven digital future.

In Chap. 4, we will present a case study featuring a smart car ecosystem developed on the proposed Web3 framework. This integration unlocks a collaborative value loop where users, infrastructure operators, developers, and policymakers co-create services and optimize mobility systems, benefiting from a transparent, user-driven data economy in an equitable manner. By decentralizing data governance and aligning incentives with societal goals, platforms like *AutoQDrive*, a Web3-empowered mobility platform, redefine the foundations of trust, participation, and innovation in digital cities. This approach is underpinned by Web3's core principles of user empowerment through DIDs, decentralized governance via DAOs, and ethical data stewardship. Ultimately, by addressing challenges such as data privacy, unsustainable tokenomics, and interoperability, Web3-empowered platforms can realize their full transformative potential. The convergence of decentralized governance, tokenized incentives, and data sovereignty reimagines smart mobility as a resilient, ethical, and participatory public service, designed for a future of more intelligent, safer, and more sustainable city mobility.

In Chap. 5, we will explore how smart cities are evolving into human-centric, decentralized ecosystems through the integration of technologies such as IoT, AI, big data, and Web3. It will highlight how smart cities address urban challenges by enhancing mobility, safety, sustainability, and governance. Innovative mobility solutions, including real-time traffic management and Mobility-as-a-Service, enhance transportation efficiency, while AI-driven surveillance and emergency response systems bolster public safety. Sustainability efforts focus on smart grids, electric vehicles, and intelligent waste management. In particular, we will showcase how Web3 technologies further empower cities by enabling asset tokenization (e.g., EV chargers, solar panels), incentivizing citizen data contributions with tokens, and supporting decentralized governance via DAOs. These innovations foster greater citizen participation and value co-creation. However, the chapter also notes challenges around technical interoperability, scalability, public awareness, and regulatory clarity. Overall, it will present a vision of smart cities as inclusive, adaptive, and ethically grounded urban ecosystems, where technology and governance work together to empower individuals and communities.

In the final chapter, we will argue that the principles of Service Science and value co-creation in Web3 can be organically and synergistically realized through the integration of blockchain, the metaverse, and privacy-preserving technologies. We particularly explore how three rapidly evolving yet interconnected technologies, AI, Trusted Data Spaces (TDSs), and the metaverse, form the backbone of Web3's transformative potential, enabling large-scale innovation and value co-creation that will profoundly impact the future of smart living. What emerges is more than just a decentralized digital infrastructure; it promotes a new societal ecosystem where value is co-created by design, agency is equitably distributed, and digital experiences are rooted in ethical, inclusive, and adaptive intelligence.

In this book, we articulate that Web3 redefines value not as something to be extracted but as a shared outcome of collaboration, collective governance, and continuous co-evolution in the future digital world. Blockchain ensures trust, the metaverse brings immersive experiences, and privacy-preserving technologies uphold security and ethical integrity. Ultimately, the *deepservice* (Qiu, 2024b)—the next generation of intelligent services that are meticulously personalized, human-centered, ethically designed, and context-aware—makes the Web3 ecosystem profoundly human.

References

Abdul Kadir, A. F., Habibi Lashkari, A., & Daghmehchi Firoozjaei, M. (2024). Android operating system. In *Understanding cybersecurity on smartphones: Challenges, strategies, and trends* (pp. 25–42). Cham.

Alam, M. S., Siddiqui, S. T., Qidwai, K. A., Aftab, A., Kamal, M. S., & Shahi, F. I. (2023). Evolution of wireless communication networks from 5G to 6G: Future perspective. *Radioelectronics and Communications Systems, 66*(5), 213–222.

Amazon. (2025). Amazon Company, Retrieved Aug. 5, 2025, from https://en.wikipedia.org/wiki/Amazon_(company)

Android. (2025). Android Operating System, Retrieved Aug. 5, 2025, from https://en.wikipedia.org/wiki/Android_(operating_system)

Arnold, M., Goldschmitt, M., & Rigotti, T. (2023). Dealing with information overload: A comprehensive review. *Frontiers in Psychology, 14*, 1122200.

Berghel, H. (1998). Who won the mosaic war? *Communications of the ACM, 41*(10), 13–16.

Berners-Lee, T., Cailliau, R., Groff, J. F., & Pollermann, B. (1992). World-Wide Web: The information universe. *Internet Research, 2*(1), 52–58.

Boyd, D. M., & Ellison, N. B. (2007). Social network sites: Definition, history, and scholarship. *Journal of Computer-Mediated Communication, 13*(1), 210–230.

Brin, S., & Page, L. (1998). The anatomy of a large-scale hypertextual web search engine. *Computer Networks and ISDN Systems, 30*(1–7), 107–117.

Cerf, V., & Kahn, R. (1974). A protocol for packet network intercommunication. *IEEE Transactions on Communications, 22*(5), 637–648.

Claffy, K. C., Braun, H. W., & Polyzos, G. C. (1994). Tracking long-term growth of the NSFNET. *Communications of the ACM, 37*(8), 34–45.

Firtman, M. (2013). *Programming the Mobile Web: Reaching users on iPhone, Android, BlackBerry, Windows Phone, and more*. O'Reilly Media, Inc.

Ghosh, S., & Gorai, S. (2024). *The age of decentralization: How Web3 and related technologies will change industries and our lives*. Productivity Press.

GVR. (2025). *Web 3.0 Market Size, Share & Growth*. Industry Report, 2030, Retrieved Jan. 2, 2025, from https://www.grandviewresearch.com/industry-analysis/web-3-0-market-report

Kiswani, J. H., Dascalu, S. M., & Harris, F. C., Jr. (2021). Cloud computing and its applications: A comprehensive survey. *International Journal of Computer Applications IJCA, 28*(1), 3–24.

Leiner, B. M., Cerf, V. G., Clark, D. D., Kahn, R. E., Kleinrock, L., Lynch, D. C., et al. (2009). A brief history of the Internet. *ACM SIGCOMM Computer Communication Review, 39*(5), 22–31.

McDonald, P. (Ed.). (2022). *The Routledge companion to media industries*. Routledge.

Mockapetris, P., & Dunlap, K. J. (1988). Development of the domain name system. In *Symposium proceedings on Communications architectures and protocols* (pp. 123–133).

Molla, R. (2017). *How Apples iPhone changed the world: 10 years in 10 charts*. Retrieved Aug. 5, 2025, from https://www.vox.com/2017/6/26/15821652/iphone-apple-10-year-anniversary-launch-mobile-stats-smart-phone-steve-jobs

Ofek, E., & Richardson, M. (2003). Dotcom mania: The rise and fall of internet stock prices. *The Journal of Finance, 58*(3), 1113–1137.

Palm. (2025). Retrieved Aug. 5, 2025, from https://en.wikipedia.org/wiki/Palm_OS

Qiu, G. (1996). *Modeling and control of a flexible manufacturing system using deterministic finite capacity automata*. The Pennsylvania State University.

Qiu, R. G. (2024a). Large language models: From entertainment to solutions. *Digital Transformation and Society, 3*(2), 125–126.

Qiu, R. G. (2024b). Web3 and service science: Perfectly aligned to maximize the values of all stakeholders in the digital economy. *Digital Transformation and Society, 3*(4), 337–339.

Ray, P. P. (2023). Web3: A comprehensive review on background, technologies, applications, zero-trust architectures, challenges and future directions. *Internet of Things and Cyber-Physical Systems, 3*, 213–248.

Reveron, D. E. (2024). *Evaluating network scalability of metaverse-applicable use cases*. Massachusetts Institute of Technology.

Roberts, L. G. (1978). The evolution of packet switching. *Proceedings of the IEEE, 66*(11), 1307–1313.

Statista. (2025a). *Number of internet users worldwide from 2005 to 2024*. Retrieved Aug. 12, 2025, from https://www.statista.com/statistics/273018/number-of-internet-users-worldwide/

Statista. (2025b). *Market share of mobile operating systems worldwide from 2009 to 2025, by quarter*. Retrieved Aug. 12, 2025, from https://www.statista.com/statistics/272698/global-market-share-held-by-mobile-operating-systems-since-2009

Tracy, K. W. (2012). Mobile application development experiences on Apple's iOS and Android OS. *Ieee Potentials, 31*(4), 30–34.

TSM. (2025). *Timeline of social media*. Retrieved Aug. 5, 2025, from https://en.wikipedia.org/wiki/Timeline_of_social_media

Wan, S., Lin, H., Gan, W., Chen, J., & Yu, P. S. (2024). Web3: The next internet revolution. *IEEE Internet of Things Journal, 11*(21), 34811–34825.

Wang, J., Lim, M. K., Wang, C., & Tseng, M. L. (2021). The evolution of the Internet of Things (IoT) over the past 20 years. *Computers & Industrial Engineering, 155*, 107174.

Yahoo. (2025). Retrieved Aug. 5, 2025, from https://en.wikipedia.org/wiki/Yahoo

Fundamental Technologies of Web 3.0 2

Abstract

In this chapter, we discuss that the transition from Web2 to Web3 aims to overcome the inherent limitations of the current centralized, platform-driven Internet. We continue to argue that Web3 represents the next evolutionary stage of the Internet in which a decentralized, user-centric digital ecosystem is rooted in the principles of people-centricity and value co-creation. Therefore, we explore the technologies required to drive the shift to Web3, which facilitate moving beyond platform dominance toward user-controlled digital identities and assets. We discuss its core, including Blockchain Technology that enables a trustless, transparent system through distributed ledgers, consensus mechanisms (PoW, PoS, BFT), and smart contracts that automate rules without intermediaries. Complementing this is privacy-preserving computation (PPC), which protects confidentiality via homomorphic encryption, Secure Multi-Party Computation, federated learning, differential privacy, and zero-knowledge proofs. Together, blockchain's decentralization, immutability, and security—combined with PPC's privacy guarantees—form the foundational technological stack powering Web3's decentralized digital future.

Keywords

Decentralized · User-centric · Value co-creation · Blockchain technology · Consensus mechanisms · Smart contracts · Privacy-preserving computation (PPC)

2.1 Introduction

Let us begin by revisiting what Web2 is and examining its limitations in addressing the needs of the future digital world. This understanding can guide us to set the stage for exploring how Web3 should evolve. More importantly, it can help us identify the

R. Qiu, *Value Co-Creation in Web 3.0*, Advances in Data Analytics, AI, and Smart Systems, https://doi.org/10.1007/978-3-032-12911-6_2

critical technological foundations that are required and must be developed and enhanced for the Web3 era as we depart from Web2.

Web2, widely recognized as the second generation of the Internet, is characterized by advanced computing, networking, visualization, and mobile technologies that enable interactive, user-driven experiences centered on business operations and management, content creation, and social interaction. As briefly reviewed in Chap. 1, Web2 has fueled the rise of Internet-based applications, defined by the following key features (Ray, 2023) as shown in Fig. 2.1 (SAFIT, 2025):

- Centralized Internet Platforms: Organization-centered platforms serve as hubs where users interact, conduct business, share various forms of information, and engage in social activities. Technically, these platforms collect and leverage vast amounts of user data to enhance the user experience, deliver personalized services, run targeted advertisements, and drive profits. However, the centralization of personal data often raises concerns about data privacy and security, user control, and the concentration of power.
- Rich Interactivity: Features such as real-time blogs, chats, comments, updates, and push notifications foster active user engagement and participation, creating dynamic, responsive digital experiences.
- Ease of Use: Visualization and intuitive computer-human interfaces substantially improve various user-friendly tools, enhancing accessibility and streamlining mobile applications. Most importantly, they ensure that even nontechnical users and underserved communities can easily access, navigate, and participate in the digital ecosystem.

Considering the benefits mentioned above, there is no doubt that Web2 has profoundly transformed connectivity, accessibility, and usability when compared to the first generation of the Internet. However, its reliance on service providers' platforms has created major obstacles that hinder their future growth. As briefly mentioned

Fig. 2.1 Illustration of the evolution of the Internet from Web2 to Web3

above, concerns about data privacy and security breaches, monopolistic control by a small number of tech giants and governmental agencies, and, more importantly, a lack of user autonomy have highlighted the need for a more decentralized, privacy-preserving, and transparent Internet. These shortcomings pave the way for Web3, a vision of the Internet that leverages blockchain and privacy-enhancing technologies to focus on and empower end-users (Anwar, 2018; Wan et al., 2024).

Technically and fundamentally, Web3 eliminates the need for intermediaries and centralized authorities, enabling end-users to interact directly and seamlessly. According to Ray (2023) and Wan et al. (2024), this is made possible by leveraging blockchain's immutable ledger, smart contracts, consensus mechanisms, and privacy-preserving technologies. Within the future-developed Web3 ecosystem, Web3 technologies ensure unparalleled transparency, security, and trust compared to Web2. More importantly, users are empowered with significant control over their identities and digital assets, fostering a more equitable and truly user-centered Internet. Because of its decentralized control and ownership, Web3 promises to create a secure, transparent, and equitable digital landscape where the future's data "oil" can flow optimally, benefiting not only service providers but also end-users. True to its promise, Web3, as shown in Fig. 2.1, when designed correctly and implemented properly, ensures that end-users have absolute sovereignty over their data and all their interactions and cyber behaviors.

Web3 is not just a technological upgrade layered on top of Web2; it fundamentally embodies the value co-creation perspective promoted in this book. It represents the next stage of the Internet's evolution: a decentralized ecosystem powered by technologies, such as blockchain, privacy-preserving computation, data science, and artificial intelligence (AI), and guided by two key principles of Service Science, people-centricity and value co-creation (Qiu, 2014).

In the early days of Web3, there was significant confusion surrounding the distinctions between blockchain technology and Web3 itself (Anwar, 2018; Krause, 2024; Wan et al., 2024). This book clarifies these concepts by recognizing blockchain technology and privacy-preserving computation as fundamental technologies and methods that underpin and empower the Web3 ecosystem, rather than being synonymous with it. For these reasons, this book uses the terms "Web3" and "Web3 ecosystem" interchangeably to refer to the decentralized, eco-friendly future of the Internet, regardless of the specific technologies or application domains.

Focusing on the core requirements of Web3, Fig. 2.2 presents the Web3 technology stack, consolidated and revised from Anwar's framework (2018) and leading global Web3 technology firms such as SAFIT (2025).

Notably, this revised architectural model is distinct because it avoids the broad, ambiguous classification of various decentralized applications as synonymous with Web3 (Anwar, 2018; Krause, 2024). Instead, it clarifies that these applications are the functional components that enable the Web3 ecosystem. As indicated in Fig. 2.2, this layered diagram illustrates the architectural interplay between blockchain technology, privacy-preserving computation, and other cutting-edge advancements in computing, networking, and data sciences that collectively enable and support the realization of the Web3 vision (Ray, 2023; Qiu, 2024; SAFIT, 2025).

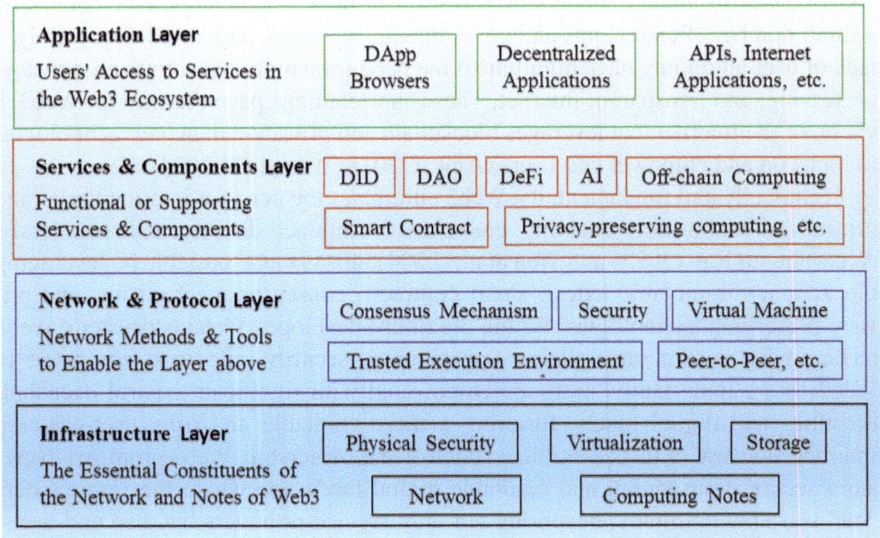

Fig. 2.2 Overview of Web3 technology stack

When examining the software aspect of Web3's hierarchical architecture, blockchain technology and privacy-preserving computation emerge as the foundational technologies of the Web3 ecosystem. These fundamental technologies enable the core functional services and components of Web3, mainly including Decentralized Identifiers (DIDs), Decentralized Autonomous Organizations (DAOs), and decentralized finance (DeFi) (Ray, 2023):

- A DID is a self-sovereign identifier designed for individuals and entities, including people, organizations, and government institutions. DIDs offer globally unique identifiers that are resolvable, highly available, and cryptographically verifiable.
- A DAO is an innovative organizational structure with no central governing authority, where participating members collectively pursue a shared goal and act in their best interests.
- DeFi is a rapidly evolving financial system built on secure, decentralized ledgers, primarily leveraging blockchain networks. By eliminating intermediaries and utilizing smart contracts, DeFi promotes and offers a diverse range of financial services and incentives in a decentralized manner. This decentralized model fosters transparency, security, and accessibility. With DeFi, individuals can finally exercise greater control over their finances and digital assets.

Drawing upon the combined capabilities of DIDs, DAOs, and DeFi, Web3 empowers users to govern and participate in decentralized applications (DApps). These applications run on a distributed network, independent of any central authority. This decentralized architecture enables seamless discovery, exchange, and use of digital assets while ensuring transparency, security, and user autonomy. In this book, we define and promote a domain-specific Web3 as a Web3 ecosystem, characterized by unique rules, domain-specific behaviors, and DApp utilities readily enabled by the collective and collaborative functions derived from DIDs, DAOs, and DeFi.

The remaining sections of this chapter will delve into the intricacies of these fundamental technologies, mainly blockchain technology and privacy-preserving computation. The next chapter will then explore the constituent functional services and components of Web3, also referred to as the pillars of Web3 in this book, in a more comprehensive manner.

2.2 Blockchain Technology

As we have discussed earlier, DApps in the Web3 ecosystem interact directly with users over a distributed network, and, more importantly, governance is collectively handled by participants rather than centralized authorities. Because of this, ensuring the immutability and transparency of participants' data and transactions on the decentralized network is crucial for the effective implementation of Web3. As is well-recognized, blockchain technology is currently the optimal technical foundation for achieving this.

2.2.1 What Is Blockchain?

Blockchain was originally presented as a decentralized, immutable ledger operating on a peer-to-peer, distributed network. It functions as a data management system, applying blocks to manage and control data and corresponding transactions over time. More specifically, at a given time, once a block is filled with the necessary information, including a timestamp, it is cryptographically hashed and linked to its preceding block, forming an unbreakable, timestamped chain of data. This chain continues to grow when new blocks are created and added in a defined logical manner. To ensure data integrity and security on the chain, a consensus mechanism with predefined business logic is adopted. Specifically, each new block must be verified and confirmed by the network before it gets added to the chain. This well-defined process creates a series of immutable, chronological records that archive all legitimate transactions on the network. Radically, this growing chain of interconnected and timestamped blocks is what widely defines the "blockchain" (Yaga et al., 2019; Ramadoss, 2022).

2.2.2 How Does Blockchain Work?

An anonymous individual or group known as "Satoshi Nakamoto" first developed and implemented blockchain technology in 2008 (Nakamoto, 2008). Satoshi

Nakamoto proposed the technology in a white paper describing the Bitcoin digital currency system, a peer-to-peer electronic cash system that uses cryptographic algorithms and distributed ledgers. It was composed of blocks linked together in a chain, with each block storing data, transactions, history, timestamps, and a cryptographic hash of the earlier blocks. This decentralized structure has since become widely adopted as a digital ledger, enabling secure and transparent records of transactions across networked computers in a decentralized, distributed, and often public fashion.

The functioning of a blockchain transaction can be explained through a simplified example (Rajasekaran et al., 2022). Consider a scenario where Alice intends to transfer specific data to John. This data could represent digital currency, such as Bitcoin, or encrypted sensitive information.

The process unfolds in several steps, as depicted in Fig. 2.3:

Step 1: Initiating the Transfer

Alice initiates the transaction by specifying the necessary details, which include:

- Sender's Information: Alice's wallet address, serving as a unique identifier for her account.
- Receiver's Information: John's wallet address, representing the destination of the transfer.

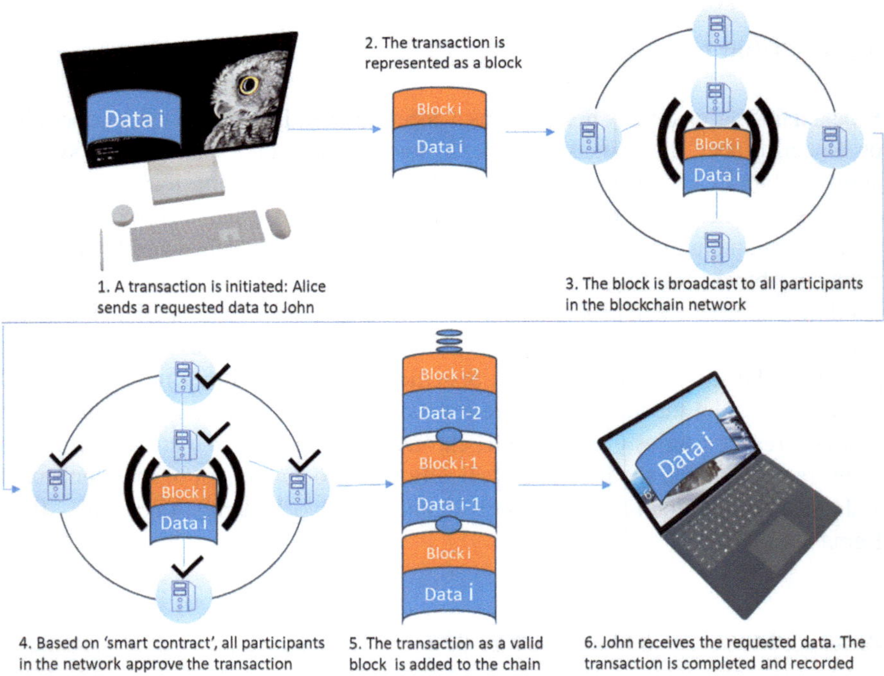

Fig. 2.3 Illustration of a typical transaction on a blockchain network

- Data Details: The specific content being transferred, such as a specified amount of digital currency or the encoded information.
- Timestamp: The precise time at which the transaction is initiated.

Step 2: Packaging the Transaction into a Block

The transaction details are consolidated and packed into a "block," a secure container designed to hold all relevant information. Additionally, a unique cryptographic identifier known as a hash value is generated and assigned to the block. This identifier guarantees the integrity and authenticity of the contents archived in the block.

Step 3: Broadcasting the Block to the Network

The newly created block is broadcast to all members of the blockchain network (referred to as participating nodes). Participants are most likely interconnected computers that store a copy of the blockchain and verify transactions, respectively.

Step 4: Verification by the Network

Each member independently verifies the contents of the block to ensure the transaction's validity. This verification process involves:

- Sender's Legitimacy: Confirming that Alice possesses sufficient digital currency and authorization to complete the transaction.
- Receiver's Validity: Ensuring that John is a recognized and valid participant within the network.

Verification is typically governed by self-executing programs (a.k.a. smart contracts) encoded with predefined rules. For example, a smart contract may require the sender to have sufficient funds and to verify the receiver's identity. Members who adhere to these rules and their authorities determine the legitimacy of the transaction. If all conditions are satisfied, the nodes approve the transaction. Otherwise, the transaction gets denied.

Step 5: Adding the Block to the Blockchain

Upon successful verification, this newly created block is appended to the chain. The blockchain functions as a digital ledger comprising a sequence of interconnected blocks, each linked to its predecessor through its hash. This structure, for example, Merkel Trees (Li et al., 2021), ensures that the blockchain is:

- Immutable: Any modification to an existing block by a participant or member requires altering all subsequent blocks and is confirmed by all other members. Tampering with the blockchain becomes virtually impossible.
- Transparent: The transaction is permanently recorded and accessible to all authorized members or participants.

Step 6: Finalizing the Transfer

Once the newly created block is appended to the chain, John receives the data or digital currency that Alice transferred. At this point, the transaction is considered complete.

2.2.3 The Essential Components of Blockchain Technology

Blockchain technology has undergone several phases of evolution since its inception in 2008. Initially, Blockchain 1.0 emerged as the foundation for cryptocurrencies, enabling decentralized, secure digital transactions. This was followed by Blockchain 2.0, marked by the introduction of smart contracts, which revolutionized automation and expanded the utility of blockchain. The Blockchain 3.0 phase saw the technology deeply integrated across a wide range of industries, demonstrating its versatility beyond finance (Li et al., 2021). Now, blockchain technology is entering its 4.0 phase, marked by powerful convergence with advances in data and computational sciences, particularly big data and AI. This integration has paved the way for truly innovative applications and significantly enhanced capabilities, moving beyond traditional cryptocurrency use cases.

Despite the rapid evolution of blockchain technology, its core functional elements or components remain relatively consistent (Li et al., 2021; Kahyaoglu & Tecim, 2024; Ramadoss, 2022; Rani & Taneja, 2024; Ressi et al., 2024; Ramachandran, 2025). These essential components of a blockchain network can be summarized and explained as follows:

- Distributed Ledger: This serves as the fundamental structure, storing shared, immutable records of all transactions across the decentralized network. It ensures transparency and auditability by maintaining a chronological, tamper-proof history of all records.
- Peer-to-Peer (P2P) Network: A decentralized network in which participants communicate with each other directly, facilitating the sharing of data, validation of transactions, and maintenance of the blockchain. This promotes decentralization and trust by eradicating the need for intermediaries on the network.
- Nodes: Participants on the network play crucial roles. Popular examples are the following:
  - Full Nodes: Store and maintain a complete copy of the blockchain.
  - Light Nodes: Store only a predefined and relevant subset of the blockchain data, reducing storage requirements.
  - Miner/Validator Nodes: Participate in the consensus mechanism by verifying the legitimacy of each transaction and adding newly created blocks to the chain.
- Consensus Mechanisms: Protocols that enable network participants to validate the legitimacy of transactions and the order of blocks on the chain. Widely adopted protocols include Proof-of-Work (PoW), Proof-of-Stake (PoS), Delegated Proof-of-Stake (DPoS), Proof-of-Authority (PoA), and Byzantine Fault Tolerance (BFT). Consensus mechanisms ensure the integrity and security of the blockchain.
- Smart Contracts: Self-executing programs embedded in the blockchain that contain predefined conditions and rules. The deployed smart contracts automatically synchronize agreements and enforce their terms, effectively removing the need for traditional intermediaries. This capability has been widely employed across

various applications, including financial and service transactions, trade, supply chain management, and polling within decentralized governance systems.
- Cryptographic Security: Robust cryptographic mechanisms are fundamental to blockchain security. Here are three widely applied cryptographies:
  - Hash Functions: These cryptographic tools ensure data integrity by creating a unique, fixed-size digital fingerprint for all blocks on the chain.
  - Public-Key Cryptography: This foundational technology enables secure communication and authentication using mathematically linked public and private key pairs.
  - Digital Signatures: Derived from public-key cryptography, digital signatures enable a robust method for participants to verify the authenticity and integrity of data and transactions on the chain.
- Native Tokens or Assets: Digital assets, often cryptocurrencies or valuable tokens, that are native to a specific blockchain. They serve various purposes, including facilitating transactions within the ecosystem, incentivizing participation (e.g., rewarding miners/validators), and fueling the development and growth of the blockchain ecosystem.

Note that consensus mechanisms and smart contracts are relatively new when compared to other elements mentioned above. Currently, they are not only unique but also indispensable components of blockchain technology. Therefore, they will be explored in greater detail in the subsections below.

2.2.3.1 Consensus Mechanisms

Consensus mechanisms play a foundational role in the realm of blockchain technology, providing a set of protocols and rules that govern agreements among distributed participating nodes on a blockchain (Yaga et al., 2019). In a blockchain ecosystem, there is no central authority to dictate the order of transactions or the state of the ledger. It is the consensus mechanisms that ensure all participants reach a shared understanding of which transactions are valid and how each newly created block should be appended to the chain.

Without consensus mechanisms, different participating nodes could possess conflicting versions of the ledger, leading to chaos and inconsistencies. For instance, in a financial blockchain, if one node accepts a transaction while others do not, the system's integrity would be compromised. Consensus mechanisms prevent such discrepancies by establishing clear criteria for consistent transaction validation across the distributed network. Due to the implementation of consensus mechanisms, the integrity of the blockchain is solidly maintained.

Moreover, these mechanisms safeguard a blockchain against a variety of potential attacks, such as double spending. In traditional digital currency systems, malicious actors could potentially use the same digital token multiple times. However, with consensus mechanisms, nodes collaborate to verify each transaction's uniqueness, making such attacks extremely difficult. This is achieved through robust cryptographic techniques and rigorous consensus-based verifications.

Furthermore, consensus mechanisms foster trust among users by ensuring that a blockchain accurately reflects all valid transactions through collective agreement among participating nodes. This trust is paramount in service sectors such as banking, investment, finance, healthcare, and international trade, where transparency and reliability are essential. Even without a central authority to oversee and enforce rules, consensus mechanisms ensure that a blockchain maintains its integrity as a trusted, verifiable source of information.

Different consensus mechanisms, when adopted, have distinct impacts on a blockchain's performance, affecting key factors such as transaction speed, energy consumption, security, and scalability. Let us examine two examples (Ramadoss, 2022; Ressi et al., 2024):

- Proof of Work (PoW): Known for its high security, PoW is computationally intensive, leading to slower transaction processing and lower throughput. PoW mechanisms, such as those used by Bitcoin, require substantial computational power. Due to its high energy consumption, PoW has raised concerns about the environmental impact of blockchain networks.
- Proof of Stake (PoS): Compared to PoW, PoS is significantly more energy-efficient and capable of processing transactions at a faster rate. By selecting validators based on the amount of cryptocurrency they "stake" as collateral, the system reduces its reliance on intensive computational power.

The choice of consensus mechanism has a profound impact on a blockchain's overall performance, influencing its ability to balance efficiency, security, and scalability while addressing environmental and operational concerns.

As blockchain technology continues to evolve, Byzantine Fault Tolerance (BFT) has emerged as a significant consensus mechanism (Zhong et al., 2023). BFT enables decentralized networks to function correctly even when some nodes are faulty or malicious, making it indispensable for ensuring the integrity, security, and reliability of distributed networks. By effectively addressing Byzantine faults, where nodes may act arbitrarily or maliciously, BFT thus ensures the blockchain's resilience and reliability.

Currently, several types of BFT consensus mechanisms have emerged in practice:

- Practical Byzantine Fault Tolerance (PBFT): PBFT is a widely used BFT that enables efficient and resource-friendly consensus. It requires a supermajority of nodes (e.g., two-thirds or more) to agree on a transaction before it is added to the blockchain. Known for its high transaction throughput, PBFT is commonly used in permissioned blockchain networks, such as Hyperledger Fabric.
- Federated Byzantine Agreement (FBA): FBA enables nodes to reach consensus by relying on a predefined set of trusted nodes (e.g., validators). Each node selects a set of trusted nodes, and consensus is reached when a sufficient number of them agree on a transaction. FBA offers high scalability and flexibility by tailoring trust relationships within a blockchain network.

- Delegated Byzantine Fault Tolerance (dBFT): A variant of PBFT, dBFT introduces a delegation mechanism where nodes vote to elect a set of delegates responsible for proposing and validating blocks. This process enhances security by reducing the risk of malicious behavior through a consensus-based election process.

Compared to earlier consensus mechanisms like PoW and PoS, BFT mechanisms offer several distinct advantages:

- Improved Transaction Speed: Mechanisms such as PBFT and dBFT reduce the time to reach consensus, enabling higher transaction throughput. This is especially beneficial for applications requiring rapid processing, such as financial transactions.
- Enhanced Scalability: BFT mechanisms enable parallel transaction processing, substantially improving the scalability of blockchain networks.
- Robust Security: By requiring the most trusted nodes to agree on the state of a blockchain, BFT mechanisms provide strong protection against malicious attacks.
- Energy Efficiency: Unlike PoW (even PoS), which requires substantial computational power, BFT mechanisms like PBFT are significantly more energy-efficient, eliminating the need for resource-intensive mining operations.

BFT has become a crucial component of modern blockchain technology, enabling enhanced security, efficiency, and scalability. By enabling networks to tolerate faulty or malicious nodes, BFT ensures the reliability and integrity of distributed systems. As blockchain technology continues to expand into diverse applications, BFT mechanisms will remain critical for achieving high performance and resilience.

In summary, consensus mechanisms are essential for enabling decentralization and trustless interactions between participants. They provide resilience against attacks, warranting the reliability and resilience of blockchain networks for mission-critical applications such as finance, healthcare, supply chain management, and identity verification. As these mechanisms continue to evolve, blockchain technology will further enhance scalability, security, and energy efficiency, laying a solid technical foundation for the broader adoption of decentralized networks across businesses and societies.

2.2.3.2 Smart Contracts

Smart contracts are self-executing programs stored on a blockchain that can autonomously execute designated agreements and obligations once predefined conditions are met (Ramadoss, 2022; Ressi et al., 2024). These programs act like digital agents, executing tasks strictly according to the rules and conditions encoded within them. As discussed earlier, they are integral to a blockchain's promise of transparency since the inception of Blockchain 2.0, as all participants involved on the network can directly view the terms and status of the contracts deployed on the blockchain.

Consider a scenario in which a research institution seeks to acquire access to a pre-priced dataset on climate change. Traditionally, this would necessitate a multi-step process:

- Negotiation and Contractual Agreement: The research team must contact the data provider, negotiate access terms, and execute a formal contract, often involving legal counsel.
- Payment Processing: The payment process would likely involve multiple internal approvals and the involvement of intermediaries such as accountants.

However, by leveraging a smart contract, this process can be significantly streamlined (Hewa et al., 2021). The smart contract, stored on a blockchain, would encapsulate the entire agreement. It would include details such as the dataset name, the predetermined price ($800), and the conditions for access (e.g., full payment triggers immediate access). Upon initiating the transaction, the research institution would simply transfer the required $800 to the designated smart contract address. Once the blockchain verifies the receipt of the full payment, the smart contract autonomously grants the institution access to the dataset. This automated process eliminates the need for intermediaries to verify payments, issue licenses, or handle administrative tasks, resulting in faster, more efficient, and cost-effective transactions (Kemmoe et al., 2020).

Based on the simplified scenario mentioned above, Fig. 2.4 illustrates the core operational steps of a smart contract within a blockchain network:

- Definition of Rules: The agreement terms, including the dataset name, the predetermined price ($800), and the access conditions (e.g., full payment), are defined in the smart contract code.
- Deployment: The smart contract code is compiled and then deployed onto the blockchain.
- Execution: Upon the research institution transferring the $800 to the designated smart contract address, the contract automatically verifies the payment.
- Recording: Once the payment is confirmed (instantly on the blockchain), the contract grants the research institution access to the dataset, and this transaction is recorded on the blockchain.

According to Hewa et al. (2021) and Kemmoe et al. (2020), the following key features of smart contracts on a blockchain can be briefly summarized:

- Automation: Smart contracts involved will execute automatically when the agreed-upon conditions are satisfied, eliminating the need for human intervention.
- Trustlessness: Participating nodes will rely solely on the code, eliminating the need to trust each other or a third party, as the blockchain ensures execution as predefined and coded.
- Immutability: Smart contracts cannot be altered once deployed to the blockchain, ensuring their integrity and security.

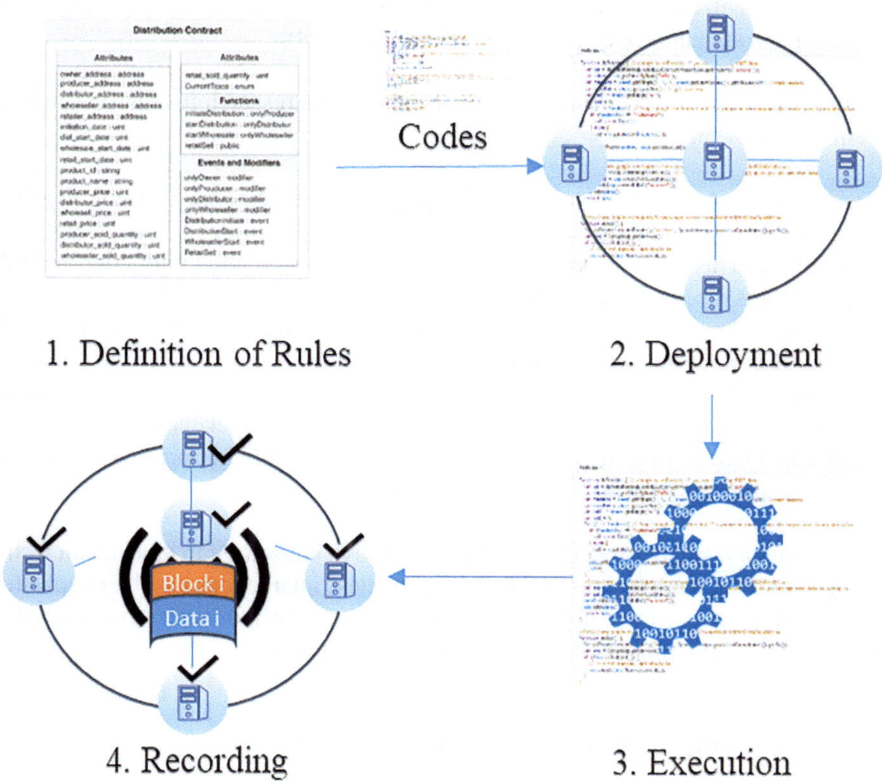

Fig. 2.4 Illustration of how a smart contract works in a blockchain network

- Transparency: The terms of smart contracts are visible to all participants on the blockchain network, promoting accountability and traceability.
- Efficiency: Smart contracts reduce delays and transaction costs associated with traditional processes by removing intermediaries and promoting automation.

Once again, smart contracts are self-executing programs stored on a blockchain that autonomously enforce predefined agreements and obligations upon the fulfillment of specified conditions. By leveraging these key features, smart contracts offer a transformative approach to fostering various blockchain-driven applications, including the following:

- Supply Chain Management: Tracking goods, ensuring authenticity, and automating payments
- Financial Services: Facilitating secure and efficient transactions, such as loans, insurance, and derivatives
- Healthcare and Data Management: Securing and sharing patient data, managing patient consent, automating insurance claims, and facilitating drug development

- Asset Tokenization: Enabling the tokenization of physical assets like individual or organizational properties, fine art, or private equity

In the context of our example mentioned above, the research institution benefits from the following:

- Automated Access: Upon successful payment, the research institution gains immediate access to the dataset without manual intervention.
- Increased Trust: The smart contract ensures access is granted only after the full payment is received, eliminating the need to trust the data provider.
- Enhanced Security: The immutability of the blockchain prevents unauthorized modification of the agreement or the dataset.
- Improved Transparency: The terms of the agreement and the transaction status can be easily verified by all parties involved within the blockchain.
- Reduced Costs: Eliminating intermediaries and streamlining processes reduces administrative overhead and associated costs.

In summary, smart contracts are a transformative innovation in blockchain technology, automating processes and fostering trustless interactions. As their capabilities continue to expand, they will continue to redefine industries and drive the widespread adoption of decentralized blockchain networks.

2.2.4 The Four Different Types of Blockchain

Campbell (2024) provides a detailed elaboration of four distinct blockchain types, each accompanied by relevant use cases. Concurrently, Kahyaoglu and Tecim (2024) present a variety of applications that effectively leverage these blockchain types—public, private, hybrid, and consortium—to meet specific domain-centric requirements. Each of these network architectures offers unique advantages and disadvantages, which are briefly summarized below.

2.2.4.1 Public Blockchain

Public blockchains are nonrestrictive, permissionless networks for anyone with an Internet connection. Any user can become an authorized node, verify transactions, access records, and participate in transactions. These blockchains are typically immutable, transparent, and open-source. They are commonly used for cryptocurrency mining and exchange, for creating auditable, for fixed records, and for applications that require transparency and trust, such as public records.

Public blockchains offer several key advantages. Their independence from any single organization ensures continuity even if the founding entity ceases to exist. Transparency is a core feature, as it allows any user to verify transactions and audit records. Many networks also incentivize users with rewards for contributing computing power, which helps secure and power the network. When security protocols are diligently followed, these networks are generally robust and secure.

However, public blockchains also present notable disadvantages. The network can be slow and may not scale out efficiently as more nodes join. Furthermore, no entity can restrict access or use of the network, which can be a limitation for specific applications. A public blockchain is also vulnerable to a "51% attack," in which an entity controlling the majority of the network's nodes can potentially alter the blockchain's transaction history.

2.2.4.2 Private Blockchain

Private blockchains operate in restrictive and permissioned settings, such as closed networks or under the control of a single entity. While they utilize peer-to-peer connections and some decentralization, they operate on a smaller scale and are typically referred to permissioned or enterprise blockchains. It is worth noting that if a private entity owns the blockchain, the technology's immutability still holds, meaning the owner cannot alter transactions.

Private blockchains are ideal for situations requiring speed, cryptographic security, and controlled access. Typical applications include managing trade secrets, auditing, healthcare, asset ownership tracking, and internal organizational voting. These blockchains are particularly suitable for businesses seeking the benefits of blockchain technology without exposing sensitive information to the public.

Private blockchains offer several key advantages. The controlling entity has the power to set permissions, security, and access levels, determining which nodes can view, modify, or add data. They are notably faster and more efficient due to their smaller size and controlled environment. Enhanced privacy is a core benefit, as third-party access to sensitive information can be restricted. They are often compared to an intranet versus the public Internet in terms of their exclusivity and control.

However, private blockchains also have several disadvantages. They face criticism for not being "true" blockchains because they lack complete decentralization Trust in the data can be weaker because centralized nodes control their validity. Smaller networks can be less secure, with risks of compromised consensus if rogue nodes act maliciously. Additionally, the use of proprietary, closed-source code limits independent auditing, reducing transparency and often leading to a lack of user anonymity.

2.2.4.3 Hybrid Blockchain

Hybrid blockchains combine the advantages of both private and public blockchains, offering a versatile solution for organizations that require a balance between confidentiality and selective public access. While data and transactions can be kept private, specific records can be verified using mechanisms such as smart contracts on a public blockchain. This architecture enables owners to secure confidential data within the network while maintaining transparency and trust.

Hybrid blockchains are ideal for applications requiring both privacy and selective transparency. Examples include the following:

- Real Estate: Operating private systems for confidential data while publicly sharing property listings
- Financial Services: Supporting operations in highly regulated industries while complying with certain regulations of aggregated public records
- Medical Records: Securely storing private medical records with patient or third-party access enabled via smart contracts
- Government: Storing citizen data privately while enabling secure inter-institutional sharing

Hybrid blockchains offer a unique blend of benefits. They operate within a closed ecosystem, which prevents external hackers from executing a "51% attack." This model effectively balances privacy with the ability to communicate with third parties. Transactions are notably fast, cost-effective, and scalable compared to public blockchains due to the smaller, controlled network. The network also protects user identities, revealing them only when they engage in a transaction.

However, hybrid blockchains have some distinct disadvantages. They offer limited transparency because much of the information is shielded from the public. Upgrading a hybrid blockchain can also be challenging as it requires coordination between the private and public components. Additionally, the potential lack of incentives for users to participate or contribute to a hybrid blockchain, as is common in public blockchains, can limit network growth and security.

2.2.4.4 Consortium Blockchain

A consortium blockchain, also known as a federated blockchain, combines features of both private and public blockchains. Unlike hybrid blockchains, it operates through collaboration among multiple organizational participants. Access is limited to these preselected participants, reducing the risk of a single entity controlling the blockchain. Consensus rules and procedures are managed by a preset group of nodes, with validator nodes initiating, receiving, and validating transactions, while participating nodes can initiate or receive them.

This type of blockchain is well-suited for applications demanding cooperation among multiple, preselected, and trusted entities. Consortium blockchains are ideal for collaborative industries where shared governance and data integrity are paramount, such as:

- Banking Services: Banks can form a consortium to validate interbank transactions and streamline complex financial operations efficiently.
- Research Organizations: It can promote and foster collaboration and secure data sharing and validation among multiple research institutions.
- Supply Chains: This model enables efficient logistics management for sensitive goods, such as food and medicine, by providing robust tracking and tracing for items in transit.

Consortium blockchains are more secure, scalable, and efficient than public blockchains due to their controlled nature. They offer access controls similar to those found in private and hybrid blockchains, ensuring that sensitive information remains within the network. The distribution of authority among multiple organizations provides a level of decentralization that mitigates the risk of a single entity abusing its power.

However, consortium blockchains are less transparent compared to public blockchains because they are not open to the public. They can be vulnerable to breaches if a member node is compromised, as this could affect the entire network. Furthermore, internal regulations or disagreements among governing members can impact network functionality, making it challenging to implement upgrades or changes.

2.3 Privacy-Preservation Computation

Privacy-preserving computation (PPC) plays a crucial role in safeguarding participants' confidentiality and privacy within the Web3 ecosystem while upholding the principles of trust and decentralization. Rooted in foundational cryptographic principles, PPC has undergone significant evolution, particularly in the context of decentralized networks. This section offers a concise overview of PPC, setting the stage for an in-depth exploration of its applications and implications within the Web3 ecosystem in subsequent chapters.

The origins of PPC can be traced back to the introduction of homomorphic encryption (HE) in the 1970s, a groundbreaking cryptographic approach that enables computations to be performed on encrypted data without decrypting it first (Acar et al., 2018). This innovation enables third parties to process sensitive data securely, preserving confidentiality. In their comprehensive survey, Acar et al. (2018) chronicle the development of HE, exploring its history, variants, and practical implementations.

In the 1980s, the emergence of decentralized networks for distributed applications necessitated the development of frameworks such as Secure Multi-Party Computation (SMPC) (Knott et al., 2021). SMPC allows multiple parties to collaboratively execute a computation on their private inputs without disclosing individual sensitive data, marking a significant milestone in the advancement of privacy-preserving technologies. In Fig. 2.5, when a user requests an aggregated report across multiple data sources, SMPC will be employed to compute it directly on encrypted data provided by contributing participants in the Web3 ecosystem. This ensures that confidential data remains secure while enabling the needed collaborative data analysis (Zhao et al., 2023).

Over the past two decades, the rise of the Internet and growing concerns about data privacy have spurred significant interest in PPC. Many stringent data privacy laws and regulations, including the General Data Protection Regulation (GDPR) and the California Consumer Privacy Act (CCPA), have further driven demand for privacy-preserving technologies. Researchers have explored practical applications of SMPC and HE in various domains, including insurance, healthcare, and finance,

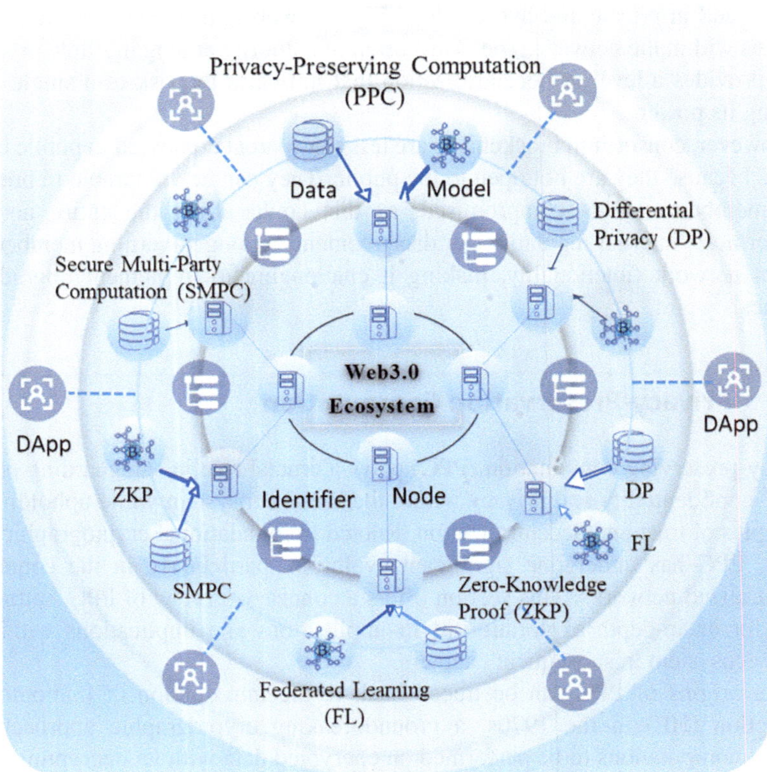

Fig. 2.5 Certain PPC's implementations in the Web3 ecosystem

as well as in blockchain networks where privacy is highly sensitive (Almashaqbeh & Solomon, 2022).

Recent advancements have expanded the scope of PPC. Federated learning (FL), a machine learning technique, enables multiple organizations to train models collaboratively without revealing their raw data, protecting data privacy (Huang et al., 2024). Differential privacy (DP), which introduces statistical noise to datasets, strikes a balance between protecting privacy and preserving data utility (Seeman & Susser, 2024). Additionally, zero-knowledge proofs (ZKPs) enable one party to prove knowledge of a fact without disclosing any sensitive information, unlocking new possibilities for secure and private collaboration (Sun et al., 2021; Zhou et al., 2024). In Fig. 2.5, when users participate in a voting process within the Web3 ecosystem, ZKPs can be used to verify their eligibility by proving they meet specific criteria (e.g., membership or affiliation) without revealing their full identity or sensitive information.

Currently, PPC presents as a rapidly evolving and promising crypto technology, transforming how data is collected, analyzed, and shared. As the field continues to advance, we can anticipate even more innovative applications of PPC, further shaping the future of privacy and security in decentralized ecosystems.

2.4 Final Remarks

Web3, often referred to as the decentralized web, represents the next stage of the Internet. By leveraging blockchain and privacy-preserving computing technologies, Web3 enables and powers DApps designed to support, encourage, and foster seamless peer-to-peer interactions. This fundamental shift eliminates intermediaries, granting users unprecedented complete control over their digital identities and information assets. Moving beyond Web2's platform-centered paradigm, Web3 promises a decentralized approach to deliver enhanced security, transparency, and a truly user-centric experience in the digital world for years to come.

In summary, the development of the foundational pillars for the Web3 ecosystem is made possible by the following critical features in the fundamental blockchain and PPC technologies comprehensively covered in this chapter:

- Decentralization: Instead of being stored on a single node, data is distributed across numerous computing nodes over a decentralized network. By eliminating centralized control, this architecture significantly improves security, enhances resilience, and strengthens resistance to censorship.
- Immutability: Once data is recorded on the decentralized network, it cannot be altered or deleted. The record becomes immutable. This fundamental property ensures the integrity, transparency, and auditability of all recorded data and transactions.
- Transparency: Every transaction recorded on a blockchain becomes publicly visible, fostering greater trust and accountability. However, the degree of privacy depends on the specific blockchain architecture and the extent to which privacy-enhancing technologies are incorporated.
- Security: Advanced cryptographic methods, such as hashing, encryption, and digital signatures, ensure the integrity of blockchain data, making it exceedingly difficult to tamper with or compromise its authenticity.

The four foundational pillars, analogous to an operating system (OS) on computers, for the Web3 ecosystem will be discussed in detail in the next chapter.

References

Acar, A., Aksu, H., Uluagac, A. S., & Conti, M. (2018). A survey on homomorphic encryption schemes: Theory and implementation. *ACM Computing Surveys (Csur), 51*(4), 1–35.

Almashaqbeh, G., & Solomon, R. (2022). Sok: Privacy-preserving computing in the blockchain era. In *2022 IEEE 7th European Symposium on Security and Privacy (EuroS&P)* (pp. 124–139). IEEE.

Anwar, H. (2018). *Web 3.0 Blockchain Technology Stack: The Comprehensive Guide*. Retrieved Jan. 10, 2025, from https://101blockchains.com/web-3-0-blockchain-technology-stack/

Campbell, C. (2024). *What are the four different types of blockchain technology*. Retrieved Jan. 10, 2025, https://www.techtarget.com/searchcio/feature/What-are- the-4-different-types-of-blockchain-technology

Hewa, T. M., Hu, Y., Liyanage, M., Kanhare, S. S., & Ylianttila, M. (2021). Survey on blockchain-based smart contracts: Technical aspects and future research. *IEEE Access, 9*, 87643–87662.

Huang, W., Ye, M., Shi, Z., Wan, G., Li, H., Du, B., & Yang, Q. (2024). Federated learning for generalization, robustness, fairness: A survey and benchmark. *IEEE Transactions on Pattern Analysis and Machine Intelligence*.

Kahyaoglu, S. B., & Tecim, V. (Eds.). (2024). *Exploring Blockchain applications: Management perspectives*. CRC Press.

Kemmoe, V. Y., Stone, W., Kim, J., Kim, D., & Son, J. (2020). Recent advances in smart contracts: A technical overview and state of the art. *IEEE Access, 8*, 117782–117801.

Knott, B., Venkataraman, S., Hannun, A., Sengupta, S., Ibrahim, M., & van der Maaten, L. (2021). Crypten: Secure multi-party computation meets machine learning. *Advances in Neural Information Processing Systems, 34*, 4961–4973.

Krause, D. (2024). *The Rise of Web3: Opportunities and Challenges*. Retrieved on Feb. 1, 2025, from https://ssrn.com/abstract=4895353

Li, W., He, M., & Haiquan, S. (2021). An overview of blockchain technology: Applications, challenges and future trends. In *2021 IEEE 11th International Conference on Electronics Information and Emergency Communication (ICEIEC) 2021 IEEE 11th International Conference on Electronics Information and Emergency Communication (ICEIEC)* (pp. 31–39). IEEE.

Nakamoto, S. (2008). *Bitcoin: A peer-to-peer electronic cash system. Satoshi Nakamoto*.

Qiu, R. G. (2014). *Service science: The foundations of service engineering and management*. Wiley.

Qiu, R. G. (2024). Web3 and service science: Perfectly aligned to maximize the values of all stakeholders in the digital economy. *Digital Transformation and Society, 3*(4), 337–339.

Rajasekaran, A. S., Azees, M., & Al-Turjman, F. (2022). A comprehensive survey on blockchain technology. *Sustainable Energy Technologies and Assessments, 52*, 102039.

Ramachandran, M. (2025). *Blockchain engineering: Secure, sustainable frameworks for healthcare applications*. .

Ramadoss, R. (2022). Blockchain technology: An overview. *IEEE Potentials, 41*(6), 6–12.

Rani, S., & Taneja, A. (Eds.). (2024). *WSN and IoT: An integrated approach for smart applications*. CRC Press.

Ray, P. P. (2023). Web3: A comprehensive review on background, technologies, applications, zero-trust architectures, challenges and future directions. *Internet of Things and Cyber-Physical Systems, 3*, 213–248.

Ressi, D., Romanello, R., Piazza, C., & Rossi, S. (2024). AI-enhanced blockchain technology: A review of advancements and opportunities. *Journal of Network and Computer Applications, 103858*.

SAFIT. (2025). *ChainWeaver: Weaving Chains into a Network and Sharing Connectivity*. Retrieved Jan. 10, 2025, https://docs.chainweaver.org.cn/index.html

Seeman, J., & Susser, D. (2024). Between privacy and utility: On differential privacy in theory and practice. *ACM Journal on Responsible Computing, 1*(1), 1–18.

Sun, X., Yu, F. R., Zhang, P., Sun, Z., Xie, W., & Peng, X. (2021). A survey on zero-knowledge proof in blockchain. *IEEE Network, 35*(4), 198–205.

Wan, S., Lin, H., Gan, W., Chen, J., & Philip, S. Y. (2024). Web3: The next internet revolution. *IEEE Internet of Things Journal, 11*(21), 34811–34825.

Yaga, D., Mell, P., Roby, N., & Scarfone, K. (2019). *Blockchain technology overview*. arXiv preprint arXiv:1906.11078.

Zhao, R., Goel, N., Agrawal, N., Zhao, J., Stein, J., Verborgh, R., et al. (2023). *Libertas: Privacy-preserving computation for decentralised personal data stores*. arXiv preprint arXiv:2309.16365

Zhong, W., Yang, C., Liang, W., Cai, J., Chen, L., Liao, J., & Xiong, N. (2023). Byzantine fault-tolerant consensus algorithms: A survey. *Electronics, 12*(18), 3801.

Zhou, L., Diro, A., Saini, A., Kaisar, S., & Hiep, P. C. (2024). Leveraging zero knowledge proofs for blockchain-based identity sharing: A survey of advancements, challenges and opportunities. *Journal of Information Security and Applications, 80*, 103678.

The Pillars of Web 3.0

3

Abstract

In this chapter, we discuss that the decentralized vision of Web3 demands a fundamental re-architecture of operational and technical paradigms. We argue that the principles of people-centricity and value co-creation in Service Science serve as guiding tenets for the development of Web3. Leveraging new Internet architectures and advanced computing technologies becomes essential for developing robust decentralized ecosystems and streamlining the deployment of the Web3 ecosystem. Therefore, in detail, we propose a Web3 technology stack and present "The Pillars of Web 3," including Decentralized Autonomous Organizations (DAOs), Decentralized Identifiers (DIDs), decentralized finance (DeFi), and decentralized applications (DApps). In other words, we identify and analyze the core building blocks necessary to develop and enable a digital future centered on enhanced security, transparency, user privacy, and self-autonomy.

Keywords

Pillars of Web3 · Building blocks · Decentralized autonomous organizations (DAOs) · Decentralized identifiers (DIDs) · Decentralized finance (DeFi) · Decentralized applications (DApps)

3.1 Introduction

In Chap. 2, we articulated that the Web3 era is coming. As we move beyond the platform-centric Web2 era, we envision a Web3-powered decentralized digital future where the Internet is driven by innovative incentives and enhanced security, transparency, user privacy, and self-autonomy. This transformative shift requires a new operational paradigm for computing and networking to reimagine the digital landscape of tomorrow. From a business perspective, the Web3 ecosystem (Fig. 3.1)

R. Qiu, *Value Co-Creation in Web 3.0*, Advances in Data Analytics, AI, and Smart Systems, https://doi.org/10.1007/978-3-032-12911-6_3

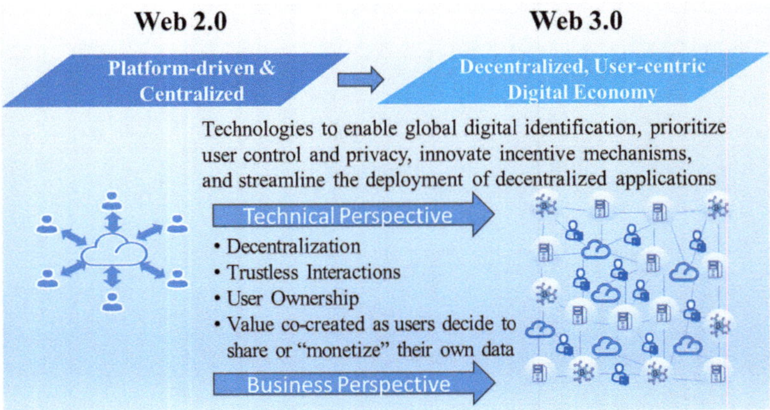

Fig. 3.1 The business and technical perspectives of the evolution of the Internet

must diverge from Web2 and adopt innovative operational models to drive digital transformation across industries and society.

As discussed in Chap. 2, the ongoing evolution of the Internet is characterized by decentralized ecosystems that have been developed, aligned with two key principles of Service Science: people-centricity and value co-creation (Qiu, 2014; Qiu, 2024), and powered by rapidly advancing computing and networking technologies. To overcome the technical shortcomings inherited in Web2, we must harness new Internet architectures and advanced computing technologies to enable global digital identification, prioritize user control and privacy, innovate incentive mechanisms, streamline the deployment of decentralized applications, and facilitate the development of the Web3 ecosystem that maximizes the value of both users and businesses (Fig. 3.1).

Achieving this vision depends on the continuous advancement and evolution of foundational technologies that underpin the Web3 ecosystem (Anwar, 2018; BTH, 2024). Four key functional components serve as the essential pillars or building blocks in Web3: Decentralized Autonomous Organizations (DAOs), Decentralized Identifiers (DIDs), decentralized finance (DeFi), and decentralized applications (DApps), as shown in Fig. 3.2. These components are the core of the fundamental Web3 technology stack, which establishes the Web3 foundational framework as indicated in Fig. 2.2 in Chap. 2.

Although a circular-layer view is presented in Fig. 3.2, all computing machines are fully decentralized across a distributed network, regardless of their roles. We use the typical layout of a computer operating system to conveniently visualize these pillars as building blocks, layer by layer. Figure 3.2 illustrates the overarching framework of the foundational technologies that underpin Web3's decentralization, user data privacy and ownership, and value co-creation innovation. Enabling this framework to leverage Web3's full potential is crucial to meticulously architecting and developing a thriving Web3 ecosystem that delivers maximum benefits to users, businesses, and society (Qiu, 2024; SAFIT, 2025).

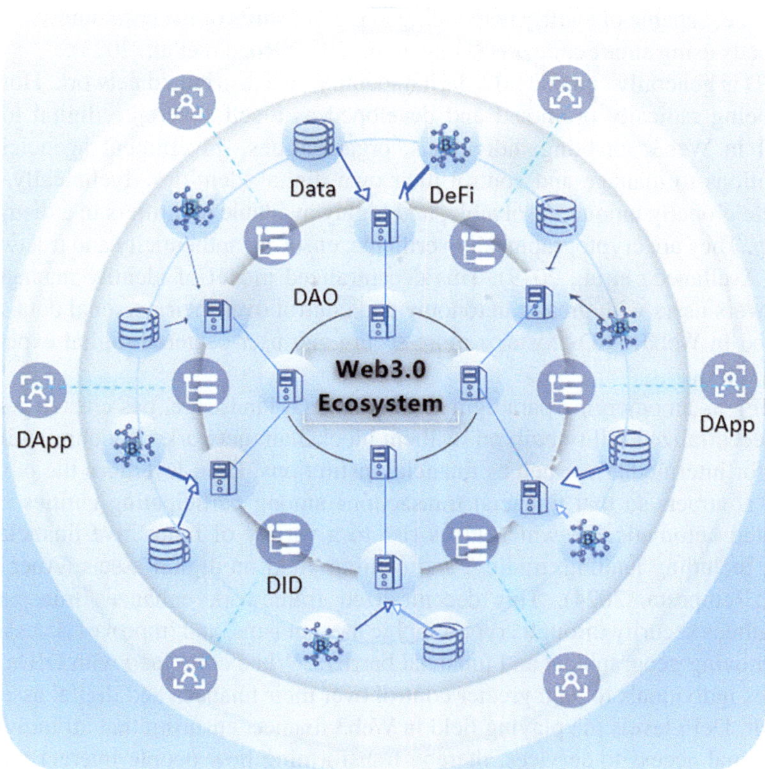

Fig. 3.2 Overview of the foundational pillars of the Web3 ecosystem

To ensure that decentralization is well supported, there should be no centralized governing authority on a distributed network. DAO becomes a groundbreaking organizational application structure that operates without centralized authority (BTH, 2024). Technically, DAOs rely on blockchain networks and smart contracts to facilitate decentralized decision-making in their operations (Ray, 2023). In general, DAO members, drawn from diverse needs and settings, unite around their common business and operational goal. Their incentives are aligned with the success and growth of their living ecosystem, aiming to foster a collaborative and self-sustaining operational framework that optimally delivers their diverse values.

A DAO member has a possessive interest in its success, whether through token ownership or active contributions to the collective effort. This decentralized approach to decision-making fosters a more democratic and inclusive organization, preventing power from being concentrated in a central authority (Surve & Khandelwal, 2023). Additionally, the transparency inherent in DAOs, where all completed transactions and actions (e.g., voting and decision-making) are recorded on a decentralized network, ensures accountability, allowing members to independently verify these transactions and actions. Furthermore, DAOs are highly adaptable, capable of swiftly responding to market shifts or the community's changing needs using smart contracts (Ding et al., 2022; Rikken et al., 2023).

DID is generally considered a digital identity on a distributed network. However, it is being radically promoted and developed as a self-sovereign digital identity model in Web3, enabling individuals, organizations, government agencies, and institutions to manage and control their own digital identities. Technically, DIDs provide globally unique, resolvable, and highly available identifiers in a distributed setting. They are cryptographically verifiable, ensuring authenticity and trustworthiness (Avellaneda et al., 2019). This decentralized model of identity management empowers users with greater autonomy and control over their personal data. When adopted in Web3, DIDs foster a more secure and user-centered digital experience (Ray, 2023).

DeFi, as an emerging paradigm in the financial landscape, has created a secure and decentralized utility built on resilient blockchain networks. By eliminating the need for intermediaries such as financial institutions, DeFi leverages the power of smart contracts so that financial transactions among participating entities can be executed automatically, which gives rise to a variety of innovative financial services, including lending, trading, and earning yield on digital assets (Auer et al., 2024; Tempesta, 2024). This decentralized framework enhances transparency, strengthens security through cryptographic mechanisms, and improves accessibility by removing geographical and financial barriers. When combined with DIDs, DeFi enables individuals to gain greater control over their finances and digital assets. As a result, DeFi levels the playing field in Web3 finance, ensuring that all individuals have equal access to services, thereby transforming how people interact with the financial world (Schär, 2021).

By integrating the collective power of DAOs, DIDs, and DeFi, DApps—the user interfaces within the Web3 ecosystem—function in a fully decentralized manner across the distributed network. By leveraging DAO's decentralized authority, these applications are collectively and coordinately managed by their end-users, promoting and encouraging a genuinely democratic and community-driven approach to the development, operation, and management of applications in Web3 (Zheng et al., 2023). Consequently, DApps in Web3 promote and promise a more equitable, accessible, and user-centric digital experience (Ray, 2023; Ghosh et al., 2024). Through the collective power of these four pillars, the empowered decentralized framework enhances transparency, security, and trustlessness, enabling all stakeholders and participants to seamlessly discover, exchange, and utilize digital assets without relying on intermediaries.

Web3 is still in its early stages of development, yet it presents a dynamic and rapidly evolving technological and commercial landscape. With its user-centric, data-driven, and value co-creation vision for the future (Qiu, 2024), Web3 aims to redefine digital interactions by fostering an open, decentralized Internet. Although the Web3 ecosystem continues to expand and transform, its foundational objective is expected to remain the same in the foreseeable future. Therefore, as technologies continue to advance, to fully unlock Web3's potential and strongly comply with the principles of decentralization, user data privacy, and ownership, as well as value co-creation innovation, its essential functional pillars must evolve in a cohesive manner. Ultimately, a Web3 ecosystem built on these four pillars can operate in a coordinated, collaborative manner to meet the demands of the future Internet as highlighted in Fig. 3.1.

Before we dive into the full discussion of those foundational services required in Web3, we would like to clarify some words that are used interchangeably in this book. These words include but are not limited to users, individuals, stakeholders, participants, nodes, elements, and members. When we discuss a Web3 ecosystem, those terms most likely mean the same thing when used to refer to participants in the ecosystem, regardless of their individual roles. Of course, they will mean differently only when their roles are spelled out in a specific deliberation.

The following sections will delve deeper into these foundational pillars of the Web3 ecosystem.

3.2 Decentralized Autonomous Organizations (DAOs)

Before the rise of DAOs, decentralized networks initially operated under centralized organizational structures, where decentralization in operations was typically achieved through hard-coded and proprietary approaches. With the rapid advancement of blockchain technology, decentralized networks can rely on consensus mechanisms, smart contracts, and other trustless protocols to ensure effective and distributed governance and control in their operations (Anwar, 2018; Yaga et al., 2019). Given that these recently advanced technologies have removed reliance on a central authority and enable decentralized operations and management, network participants can be actively engaged in decision-making and governance without needing to consult the centralized authority. Promisingly, this decentralized model has become a cornerstone of blockchain-powered applications, promoting and supporting their operational transparency, data security, and community-driven development and operation.

As Web3 continues to evolve, DAOs have become a crucial and foundational component, serving as a key building block of the Web3 ecosystem. DAOs, technically leveraging smart contracts and consensus mechanisms, enable trustless and transparent decision-making, without the need for traditional hierarchical or centralized management on a distributed network. In this book, we strongly argue that DAOs have promoted and pioneered an automated, self-governing model that aligns seamlessly with Web3 principles, facilitating the development of a truly equitable, efficient, and robust digital ecosystem in both the current and future digital economies.

Below is a detailed explanation of how DAOs automate participation and governance on a decentralized network. By referencing the recent publications on DAOs, we primarily focus on two key areas: the core functions and the future evolution of DAOs (Wang et al., 2019; Wright, 2021; Qin et al., 2023; Bellavitis et al., 2023; Rikken et al., 2023; Han et al., 2025).

3.2.1 Core Functions and Future Evolution of DAOs

3.2.1.1 Decentralized Governance and Enhanced Decision-Making Mechanisms

In a decentralized network, DAOs enable members to collaboratively define and develop rules, policies, and operational decisions through distributed voting mechanisms. Smart contracts are utilized to automatically enforce these rules and policies, executing decisions that deliver transparency, security, and fairness. Because centralized authorities are eliminated, DAOs inherently facilitate not only minimizing human bias but also reducing the risk of manipulation, enabling trustless, decentralized decision-making, and promoting governance automation.

Despite their promises, current DAO governance models face significant challenges, including voting inefficiencies and the risk of the "tyranny of the majority," where dominant groups may override the interests of minorities. To address these issues, further advancements in quadratic voting, reputation-based governance, and decentralized dispute resolution will help establish more inclusive, equitable, and scalable governance frameworks (Han et al., 2025). These improvements will enhance DAOs' ability to manage complex decision-making processes while ensuring fairness, inclusiveness, and long-term sustainability without compromising the promised benefits.

3.2.1.2 Resource Management and Improved Security

Resources can be physical or digital. If a physical asset is involved, it will be digitalized when managed on a decentralized network. As for resource management, DAOs can oversee various shared digital assets, including treasuries, non-fungible tokens (NFTs), digital resources, and intellectual property. With smart contracts, the allocation and distribution of these resources can be automated, minimizing the need for intermediaries and ensuring transparent, efficient, and secure asset management. As a result, by leveraging decentralized, community-driven governance, DAOs can facilitate automated resource management, thereby promoting co-creation and innovation.

Despite improved resource management efficiency, smart contracts used in DAO development and operations may pose security risks (Wang et al., 2019). For example, when smart contracts are used to enable DAO features on a decentralized network without appropriate security measures, they can lead to vulnerabilities and financial losses. Hence, strengthening contract security, risk mitigation, and operational resilience is critical for DAOs to scale effectively in Web3. Without a doubt, further research can be conducted to advance the following three areas for improved security when smart contracts are implemented in DAOs (Huang et al., 2019; Rikken et al., 2023). The first one is formal verification, which ensures the smart contract code is mathematically proven to be secure. The second one concerns advanced auditing mechanisms that implement more robust security audits to detect and prevent potential exploits. Finally, optimization algorithms can be developed to enhance transaction efficiency, reduce costs, and improve scalability.

By integrating these innovations, DAOs will be able to securely deploy and manage resources at scale, mitigate risks, and optimize operational efficiency, solidifying their role as a crucial foundational component in the Web3 ecosystem.

3.2.1.3 Community-Driven Operations and Incentive Optimization

DAOs revolutionize community engagement by empowering members to propose, discuss, and vote on initiatives, promoting a strong sense of ownership and inclusiveness (Tang et al., 2025). Through smart contracts, these community-driven decisions are automated and executed transparently, ensuring the organization operates in alignment with its members' collective will while reducing reliance on centralized authority.

To promote value co-creation innovation and sustain active participation and long-term engagement, DAOs must implement more effective and balanced incentive mechanisms. The following three approaches require much attention from the research community worldwide. The first one focuses on applying monetary rewards, leveraging token-based incentives to encourage contributions and governance participation from all members. The second one is related to reputation-based contributions, which essentially promote nontransferable reputation systems to recognize members' expertise and long-term commitment. Finally, forms of nonmonetary incentives can be developed to offer members recognition, voting power, and access to exclusive DAO privileges.

Future research in behavioral economics, stakeholder models, and gamified participation strategies will be essential for refining and further developing these mechanisms for adoption in Web3. Ultimately, by integrating innovative engagement models, DAOs can optimize participation, enhance inclusive governance, and ensure long-term sustainability within the Web3 ecosystem.

3.2.1.4 Transparency, Interoperability, and Legal Recognition

Within a DAO, all decisions, financial transactions, and operational activities on the network are recorded and stored appropriately. By leveraging blockchain technology, these data and actions become publicly visible (if desired in design) and cannot be altered. These records thus become transparent and immutable. Furthermore, as members can independently verify these actions, it substantially reduces the likelihood of fraud or data manipulation. All in all, DAOs enhance the transparency, trust, and accountability of a Web3 ecosystem on a decentralized network.

As DAOs in the Web3 ecosystem continue to evolve, interoperability between DAOs becomes increasingly essential. The development of standardized protocols for cross-DAO collaboration and resource sharing will facilitate the creation of a more interconnected and cohesive function pillar. Different DAOs can interact seamlessly with each other.

Furthermore, because DAOs are still relatively new, their legal status is not yet appropriately defined in various domains and countries. Addressing this challenge requires the cooperation of researchers, policymakers, and blockchain and Web3 experts. They should develop regulatory frameworks that address critical issues relevant to their specific domains and countries (Wright, 2021), such as liability, taxation, and compliance. Widely accepted regulatory frameworks will play a crucial role in facilitating the broader adoption of DAOs in the Web3 ecosystem.

3.2.1.5 A Brief Summary

By integrating these core functions with future technological advancements, DAOs will be poised to evolve into more resilient, efficient, and sustainable organizational structures. The automation of governance, resource management, community engagement, and transparency through smart contracts and blockchain technology strengthens DAOs' role as a core pillar of Web3. As DAOs continue to innovate in the digital economy, they will promote and enable more inclusive, secure, and globally accessible participation from more users than ever before, transforming how members collaborate, co-design, and co-create value in the Web3 ecosystem.

3.2.2 Existing DAO Examples

Recently, DAOs have gained widespread recognition as a prominent application of Web3 in practice (Kozhan & Viswanath-Natraj, 2022; Krause, 2024), although, as a Web3 application, they conceptually differ from the proposition articulated in this book. Interestingly, some researchers have already proposed integrating DAOs with Web3 to develop decentralized applications tailored to specific market needs (Ding et al., 2022). By the end of 2024, notable DAO implementations had been widely adopted across decentralized cryptocurrency platforms, with MakerDAO serving as a key example in powering stablecoin governance (Brennecke et al., 2022) and Uniswap DAO playing a crucial role in enabling decentralized exchange governance (Kim, 2021; Vasileva, 2025).

3.2.2.1 MakerDAO

MakerDAO governs the Maker Protocol, a pioneering DeFi platform that issues DAI, a prominent algorithmic stablecoin tied to the US dollar. The protocol facilitates the creation of DAI by allowing users to deposit cryptocurrencies (collateral) into smart contract logic. In exchange, users receive DAI, a stablecoin designed to maintain a stable price relative to the US dollar (Brennecke et al., 2022; Kozhan & Viswanath-Natraj, 2022).

According to Brennecke et al. (2022) and Kozhan and Viswanath-Natraj (2022), the significance of MakerDAO can be briefly summarized as follows:

- Pioneering Decentralized Finance: MakerDAO is a seminal example of DeFi, demonstrating the feasibility of creating and managing complex financial systems without the need for intermediaries.
- Innovative Stablecoin Mechanism: The DAI stablecoin, underpinned by the Maker Protocol, has significantly contributed to the growth of the DeFi ecosystem by providing a stable and reliable store of value and a medium of exchange within the DeFi space.

- Decentralized Governance Model: MakerDAO demonstrates a robust, evolving governance model. Members as token holders are directly involved in all crucial decision-making processes, including voting on the inclusion of new collateral types, adjusting stability fees (interest rates charged on collateral), and approving protocol changes. This democratic approach encourages and supports the community in adapting to the protocol's future, aiming to ensure its long-term sustainability and viability.
- Resilience and Adaptability: The Maker Protocol has demonstrated resilience in navigating market volatility and adapting to evolving market conditions. Continuous improvements and upgrades to the protocol have enhanced its stability, security, and efficiency.

3.2.2.2 Uniswap DAO

Uniswap DAO governs the Uniswap decentralized exchange (DEX), a leading automated market maker (AMM) that facilitates the decentralized trading of cryptocurrencies. The DAO empowers UNI token holders, representing stakeholders in the Uniswap ecosystem, to participate in the decision-making processes that shape the protocol's future (Kim, 2021; Adams, 2024).

According to Kim (2021) and Adams (2024), the significance of Uniswap DAO can be briefly summarized as follows:

- Pioneering Decentralized Exchange Governance: Uniswap DAO is a prominent example of how decentralized governance can be effectively applied to develop a large-scale and complex financial protocol and application. Promisingly, it demonstrates that the Uniswap DAO is feasible for promoting value co-creation and community-driven governance in decentralized finance applications.
- Community Empowerment: The Uniswap DAO empowers UNI token holders by granting them voting rights. By taking control of their voting rights, the Uniswap DAO allows them to directly influence the development and direction of the DEX. This essentially fosters a sense of ownership as the Uniswap DAO encourages active participation from all of them.
- Efficient and Transparent Governance: Governance automation mechanisms are well implemented in the Uniswap DAO, promoting and supporting transparent, efficient decision-making. Furthermore, proposals from any UNI token holder are publicly available, and voting results are transparently recorded on the blockchain, ensuring accountability and trust on the DEX network.
- Innovation and Adaptability: The Uniswap DAO structure and protocols are adaptable, evolving in response to market conditions and technological advancements. Because value co-creation and community-driven proposals and upgrades are consistently encouraged and supported, continuous improvement and innovation become possible within the evolving Uniswap ecosystem.
- Driving DeFi Innovation: The success of the Uniswap DAO has inspired other decentralized finance projects. The wide adoption of decentralized governance models further drives innovation in the DeFi space. In addition, through continu-

ous innovation and improvement, this has also spurred the development of decentralized non-finance applications.

3.2.3 The Essential Role of DAOs in the Web3 Ecosystem

After a comprehensive exploration of DAOs, we understand that DAOs are innovative organizational structures that operate technically through a combination of smart contracts, decentralized governance, and token-based incentive schemes. They automate various aspects of organizational functions, supporting treasury management, voting, reward distribution, membership management, and other utilities related to organizational resources and behavior. In Web3, DAOs are promoted to operate on blockchains, which inherently ensures transparency, trustlessness, and immutability as discussed earlier. They thus facilitate global collaboration, automate compliance, and enable dynamic adaptation to changing circumstances over time.

As argued earlier, to effectively support the ongoing evolution of the Web3 ecosystem, DAOs have emerged as a core component. Specifically, we argue that, much like the operational systems in computing machines, DAOs will play a crucial role within the Web3 operating system, promoting and enabling the automation of governance, resource management, and value co-creation across decentralized networks.

To facilitate the development of a Web3 ecosystem, DAO core functions can be categorized into distinct utility groups. These utility groups can be readily tailored, assembled, and deployed for specific domain-driven application needs. Below, we briefly outline six distinct utility groups within the DAO's multifunctional framework by summarizing all the explorations in this section on DAOs and developments that are largely discussed in the literature (Bellavitis et al., 2023; BTH, 2024; Krause, 2024; Han et al., 2025). Without a doubt, these utilities will manage and facilitate efficient, secure, and inclusive operations while fostering participation, collaboration, and value co-creation innovation in Web3.

3.2.3.1 Utility Group 1: Decentralized Governance and Transparency

This utility group embodies the core principles of DAOs, including decentralized decision-making and full transparency, ensuring that governance structures remain inclusive, trustless, and verifiable. By eliminating central authorities, DAOs create an open, self-regulated environment where all members are free to voice their opinions and vote in decision-making, and all actions remain auditable on the distributed network. Key sub-functions include the following:

- Decentralized Governance: Implementing token-based voting systems and layered governance structures to facilitate collective decision-making
- Transparency and Trustless Operations: Ensuring that all records are immutable and operations are trustless, allowing full visibility and accountability

By incorporating decentralized governance and transparent operations into their core structure, DAOs promote fairness, inclusiveness, and trust, thereby ensuring long-term sustainability and resilience within the Web3 ecosystem.

3.2.3.2 Utility Group 2: Automation and Efficiency

This utility group focuses on leveraging automation to enhance operational efficiency, minimize human intervention, and ensure compliance within DAOs. By integrating smart contracts and decentralized protocols, DAOs can streamline governance, optimize resource management, and maintain transparency, reducing inefficiencies commonly associated with traditional, centralized organizational structures. Key sub-functions include the following:

- Automation through Smart Contracts: Automating critical processes such as treasury management, voting, reward distribution, and membership management
- Operational Efficiency: Streamlining routine tasks and ensuring scalability for growing DAOs
- Automated Compliance and Reporting: Ensuring adherence to regulations and providing transparent reporting mechanisms without manual intervention

By integrating these automation-driven efficiencies, DAOs can significantly reduce operational friction, enhance governance responsiveness, and establish a more sustainable and scalable decentralized ecosystem.

3.2.3.3 Function Group 3: Security and Trust

Security is crucial for DAOs, ensuring they are resilient against vulnerabilities, attacks, and malicious actors while upholding the principles of decentralization and trustlessness. This group ensures the protection of assets, transactions, and participants in the DAO applications (Ray, 2023). Key sub-functions include the following:

- Automated Security Measures: Implements advanced security tools, including multi-signature wallets, audited smart contracts, and formal verification of decentralized identities
- Real-Time Monitoring and Threat Detection: Leverages decentralized security frameworks for continuous monitoring and threat detection

3.2.3.4 Utility Group 4: Incentive and Reward Systems

This utility group is crucial for aligning participants' interests with the DAO's goals, promoting long-term engagement, motivation, and active participation. By implementing well-structured incentives and reward mechanisms, DAOs can sustain participation, encourage collaboration, and drive innovation within their ecosystems. Key sub-functions include the following:

- Automated Incentive Structures: Designing token economies and reward distribution systems that encourage active participation

- Automated Value Distribution: Ensuring fair compensation through token-based rewards to motivate contributions and maintain organizational health

By refining incentive structures and enhancing automation in value distribution, DAOs can create a self-sustaining ecosystem where members are fairly rewarded, deeply engaged, and actively contributing to the organization's growth and success.

3.2.3.5 Utility Group 5: Global Collaboration and Accessibility

This utility group highlights DAO's ability to facilitate seamless global collaboration and participation, breaking down geographical barriers. Key sub-functions include the following:

- Global and Permissionless Participation: Enables cross-border transactions and participation without geographic limitations
- Cross-DAO Collaboration: Fosters interaction and cooperation between different DAOs

This group empowers DAOs to transcend geographical limitations and establish a truly global, decentralized ecosystem where every individual can contribute to the network's success.

3.2.3.6 Utility Group 6: Dynamic Adaptability and Evolution

DAOs are inherently designed to evolve and adapt over time, enabling responsive changes driven by community input and ecosystem shifts. This adaptability ensures that DAOs remain resilient and able to adjust to new challenges, opportunities, and advancements. Key sub-functions include the following:

- Dynamic Adaptation: Facilitates the evolution of rules, governance structures, and operational protocols, enabling continuous improvement and adaptation
- Continuous Feedback Loops: Embeds mechanisms for ongoing community feedback, allowing members to actively participate in shaping the DAO's future

This group ensures that DAOs can adapt to both internal changes, such as shifts in membership or priorities, and external factors, including technological developments or changes in the regulatory landscape, thereby maintaining long-term sustainability and innovation.

In summary, the Web3 ecosystem can harness the six core DAO utility groups to build flexible governance structures. By selectively activating these functions across different utility groups, organizations can tailor DAOs to meet their evolving needs and diverse use cases across various industries and societies. In this book, once again, rather than being treated as domain-specific applications, DAOs are more accurately defined as a core functional component, a pillar or building block, that empowers and facilitates the development, operations, and management of the entire Web3 ecosystem.

3.3 Decentralized Identifiers (DIDs)

The online identities of network participants have evolved through four key stages since the inception of the Internet. These stages are described as centralized identity, federated identity, user-centric identity, and self-sovereign identity, each marking a significant shift in participants' control and ownership of their identities on distributed networks (Avellaneda et al., 2019). This progression truly reflects their gradual, fundamental shift in power, clearly indicating a movement away from reliance on centralized authorities and toward a decentralized identity model. Users, whether individuals or organizations, as participants, have gained increasing control over their own digital representations across various networks.

Currently, the emerging Decentralized Identifiers (DIDs) are at the cutting edge, serving as a revolutionary type of cryptographically verifiable identifier. DIDs represent a paradigm shift, truly enabling self-sovereign digital identities as articulated above. Unlike traditional identifiers, which are tied to and managed by centralized intermediaries such as governments or corporations, DIDs provide tools and methods to enable individual participants to independently create and manage their own identities (Mazzocca et al., 2024). This eliminates the need to rely on central authorities, granting users complete autonomy over their digital presence in the online world.

As of 2025, this decentralized approach has formed the bedrock of truly decentralized identity systems, promising enhanced privacy, robust security, and seamless interoperability across diverse digital ecosystems (Avellaneda et al., 2019; Mazzocca et al., 2024). This is what is expected in the Web3 ecosystem: users control their digital credentials, deciding who has access to what information, without fear of data breaches or centralized censorship.

For widespread adoption in practice, DIDs were standardized by the World Wide Web Consortium (W3C) in 2019. In 2022, they were meticulously designed to be globally unique, resolvable, and tamper-resistant (Sporny et al., 2022). This trifecta of attributes enables trustless authentication and secure interactions, particularly in the emerging Web3 landscape. Because DIDs leverage a decentralized ledger, they provide verifiable proof of identity ownership without the need for intermediaries or third parties to confirm an individual's credentials. As a result, this emerging DIDs approach can not only streamline authentication processes but also facilitate the enablement of a more secure, transparent, and efficient Web3 digital environment (Tempesta, 2024).

3.3.1 DIDs Use Cases

DIDs are a transformative technology that promotes verifiable, self-sovereign digital identities (Tempesta, 2024). Technically, DIDs are designed to be globally unique, cryptographically secure, and independent of centralized authorities. To promote decentralized digital identity technology in Web3, one emerging DID platform is called Truvera[1] by Dock Labs (Dock, 2024). Below, we briefly explore a few general use cases for DIDs, highlighting their versatility and potential to revolutionize identity management in the Web3 era.

3.3.1.1 Digital Identity

DIDs enable the creation of portable, self-sovereign identities, allowing users to access online services without depending on centralized identity providers. Authentication is shifting from classic usernames and passwords to cryptographic proofs, which offer a more secure method and lower the risk of data breaches and identity theft (Ray, 2023). In Web3, this supports users to seamlessly interact across DeFi platforms, DAOs, metaverses, NFT marketplaces, and more while maintaining full control over their personal data. In fact, DIDs can be fully integrated into any Web3 ecosystem.

3.3.1.2 Verifiable Credentials

DIDs facilitate the issuance, storage, and verification of digital credentials (e.g., diplomas, professional certifications, driver's licenses, deeds) in a privacy-preserving manner. A university, for example, can issue a verifiable credential to a graduate's DID, enabling them to prove their qualifications without revealing unnecessary personal details. This is especially useful in virtual decentralized job marketplaces, where credentials can be verified over an integrated blockchain without the need for intermediaries or third parties.

3.3.1.3 Decentralized Autonomous Organizations (DAOs)

As discussed earlier, DAOs can readily leverage DID-featured identity verification to ensure fair, secure, and inclusive governance. DIDs can also be applied to warrant Sybil resistance (preventing users from creating multiple fake accounts to manipulate voting). By associating blockchain-enabled reputation scores or verifiable credentials within DIDs, DAOs can implement more transparent, merit-based decision-making processes while preserving the privacy of individual members.

3.3.1.4 Decentralized Finance (DeFi)

DIDs can enhance trust and security in DeFi by leveraging blockchain-featured reputation systems and risk assessments. For example, users can link verifiable credentials (e.g., proof of creditworthiness) to their DIDs, allowing DeFi lending platforms to offer personalized loan terms without exposing sensitive personal details. This fosters a co-creation, trustless, and secure financial system. Further discussions on DeFi will be presented in the following section.

3.3.1.5 Decentralized Applications (DApps)

DIDs provide a secure, interoperable, and decentralized authentication mechanism for users interacting with blockchain-based applications. As DIDs are verifiable across distributed networks, users can use a single DID to log in, verify their assets,

[1] https://www.dock.io/.

and interact with smart contracts across multiple platforms, eliminating the need for multiple accounts for different DApps. This enhances user experience, security, and privacy, making it easier to engage with DeFi protocols, NFT platforms, and decentralized social networks, as well as other emerging Web3 applications, without relying on third-party identity providers. Note that further discussions on DApps will be presented at the end of the section.

3.3.2 An Illustrative Example of Applying DIDs in Web3

Imagine Phoebe, a Web3 enthusiast, wants to participate in a DAO without revealing unnecessary personal information (Sporny et al., 2022; Mallare et al., 2023). She leverages a DID linked to her digital wallet or DApps to achieve this. Here is how the process unfolds:

3.3.2.1 Step 1. Creation of a DID

Phoebe's digital wallet (a Web3 application) generates a unique DID and an associated cryptographic key pair. The DID is created using a DID Method (e.g., did:safit for SAFIT-based identities) and is stored securely in her wallet. The DID document, which contains her public key and service endpoints, is published on a decentralized ledger (e.g., Ethereum or IPFS).

- Technical Detail: The DID might look like this: did:safit:GL984643609845PiYAdyhkjf876549870783UhhefuiHdkf9867436
- User Experience: Phoebe does not need to reveal any personal data, such as name, address, phone number, or driver's license number, to create the DID. She simply clicks "Create Identity" in her wallet.

3.3.2.2 Step 2. DAO Membership Verification

This DAO function requires proof of governance token ownership to grant membership. Phoebe's wallet holds the required tokens, and she proves her ownership by signing a cryptographic challene with her private key. This DAO function verifies the signature using her DID's public key, confirming her eligibility without requiring knowledge of her real-world identity.

- Technical Detail: The DAO uses a smart contract to verify token ownership and DID-based authentication.
- User Experience: Phoebe clicks "Join DAO" in her wallet, and the process happens seamlessly in the background.

3.3.2.3 Step 3. Participation and Voting

Once verified, Phoebe uses her DID to participate in the DAO's governance processes. For example:

- She votes on proposals by signing her vote with her private key, ensuring the vote is tied to her DID.
- She submits proposals by attaching her DID to the proposal, proving her membership and authenticity.
- Technical Detail: Each vote or proposal is cryptographically signed and recorded on the decentralized network, ensuring tamper-proof transparency.
- User Experience: Phoebe interacts with the DAO through a user-friendly interface in her wallet, without creating new accounts or sharing personal data.

3.3.2.4 Step 4. Interoperability Across Web3 Ecosystems

Later, Phoebe wants to access a DeFi lending platform. Instead of creating a new account or undergoing a lengthy "Know Your Customer" (KYC) process, she uses the same DID to prove her identity. The DeFi platform recognizes her DID and verifies her credentials (e.g., creditworthiness or reputation) through verifiable credentials issued by the DAO or other trusted members.

- Technical Detail: The DeFi platform queries Phoebe's DID document to retrieve her public key and verify her identity. It may also verify on-chain reputation systems associated with her DID.
- User Experience: Phoebe logs in to the DeFi platform using her DID, and her credentials are instantly verified, enabling her to access services without friction.

Throughout this process, Phoebe maintains full control over her identity and personal data:

- She never shares unnecessary personal information (e.g., name, address, phone number, or driver's license number).
- Her DID is portable and can be used across multiple platforms without requiring the creation of new accounts.
- She can revoke access to her DID at any time, ensuring her identity remains under her control.

By using DIDs as a building block of Web3, Phoebe can use DApps, participate in DAOs, access DeFi platforms, and interact with other Web3 services in a secure, private, and interoperable manner. This example highlights the transformative potential of DIDs in enabling self-sovereign identity and fostering a more user-centric Internet. As Web3 evolves, DIDs are a vital technology that empowers users, such as Phoebe, to retain ownership and control of their digital identities and other sensitive personal information, aligning with the core principles of decentralization, user-centricity, community, and value co-creation through active participation.

3.3.3 The Key Characteristics of DIDs in Web3

As digital identity technology continues to evolve inexorably. DIDs stand as a pivotal innovation over the years. As DIDs continue to evolve, they will undoubtedly become a cornerstone in the construction of a more open, user-controlled, and resilient identity framework for the future of the Internet. Indeed, DIDs become essential as a key enabler of the Web3 vision, where individuals are empowered with full control over their digital lives.

As explored in this section, integrating DIDs into the Web3 ecosystem enables users to hold secure, privacy-preserving, and self-controlled identities, thereby redefining trust and authentication in decentralized digital interactions. Hereafter, we can summarize the key characteristics of DIDs necessary for the Web3 ecosystem as follows (Avellaneda et al., 2019; Ray, 2023; Mazzocca et al., 2024; Tempesta, 2024):

- Decentralization—In contrast to traditional identifiers (e.g., email addresses, usernames/passwords) controlled by centralized platforms such as gaming service providers and social media networks, DIDs are created and managed by individual participants or end-users. When incorporated into the Web3 ecosystem (i.e., as a building block or part of the Web3 OS), this inherent decentralization, embedded in DIDs, eliminates reliance on central registries, certificate authorities, or intermediaries, ensuring trustless identity creation and management in distributed applications.
- Self-Sovereignty—DIDs align well with the principles of self-sovereign identity (SSI), enabling individual participants to manage their digital identities independently. In Web3, this enables users to authenticate, interact, and transact across DApps, DAOs, DeFi, other emerging Web3 OS utility services, and various distributed networking protocols while maintaining full autonomy over their personal data and exercising consent-driven sharing for improved data protection.
- Cryptographic Verifiability—Typically, each DID is cryptographically secured with public-private key pairs, allowing the identity owner to prove control over their DIDs in every transaction when a verification is required. In Web3, this ensures secure, tamper-proof authentication, reducing fraud risks and providing the foundational decentralized trust mechanisms that are radically essential for smart contracts, DeFi platforms, and digital credentialing.
- Persistence—DIDs are designed for longevity, independent of any single authority or service provider. Unlike traditional identity systems that may become obsolete if a provider ceases operation, DIDs persist across decentralized networks, ensuring their longevity. This is critical for a highly resilient Web3 ecosystem, where identity continuity is essential for reputation and reward systems, decentralized governance, and participatory value co-creation innovation.
- Interoperability—As mentioned earlier, DIDs are built on open and standardized protocols as endorsed by W3C. Hence, DIDs can support seamless identity verification across different blockchains, decentralized identity frameworks, and Web3 ecosystems. This interoperability enhances portability, allowing users to

maintain a unified digital identity across DeFi platforms, DAOs, metaverses, and other decentralized services without being locked into any single ecosystem.

In summary, DIDs represent a paradigm shift in digital identity management on distributed networks, offering a secure, privacy-preserving, and interoperable framework for identity verification. With the enabled self-sovereign identities, DIDs have the potential to transform how individuals and organizations interact in the future, in a highly decentralized digital world, fostering trust and promoting value co-creation and innovation. In the long run, however, realizing the full potential of DIDs requires addressing technical, social, and governance challenges, including but not limited to the following:

- Scalability: Because DIDs are globally unique, efficiently managing and resolving large volumes of them in real time poses a technical challenge.
- Adoption: Widespread adoption requires recognition and acceptance from governments, organizations, and individuals across countries and continents, as well as effective strategies and tactics for integrating with existing systems.
- Privacy: While DIDs enhance privacy by reducing reliance on centralized platforms, improper implementation can still pose privacy risks.
- Governance: Decentralized systems require robust governance models to ensure fairness, security, and accountability when widely adopted across businesses, industries, and societies.

3.4 Decentralized Finance (DeFi)

Decentralized finance (DeFi) is an innovative financial technology built on secure decentralized or distributed ledgers, typically running on a blockchain-based network or similar decentralized network platform. It introduces more effective and efficient mechanisms to enable the equitable distribution of value generated by individual participants within the network, fostering transparency, accessibility, and inclusiveness in financial or incentive systems.

3.4.1 A Brief History of DeFi and Its Benefits

DeFi aims to revolutionize financial services by establishing a transparent and accessible system that eliminates the need for traditional intermediaries, such as banks. By leveraging blockchain and privacy-preserving technologies, DeFi empowers users with self-control and automates key functions, including lending, borrowing, exchanging, and value distribution, thereby enabling users to focus on providing greater control and potentially more rewarding opportunities (Auer et al., 2024). Although DeFi is still relatively new, its evolution has been broadly categorized into four key phases, each building upon the innovations of its predecessors.

3.4.1.1 Foundational Building Blocks (2008–2016)

The introduction of Bitcoin in 2008 marked the beginning of DeFi (Nakamoto, 2008). Bitcoin demonstrated the feasibility of a decentralized, peer-to-peer electronic cash system, laying the foundation for the broader concept of DeFi. This initial phase focused primarily on the creation and exchange of digital currency independent of traditional financial institutions. Later, the advent of Ethereum marked a pivotal moment in the utilization of smart contracts (Buterin, 2014). It is worth mentioning that the inclusion of smart contracts provided the necessary infrastructure to develop more complex DApps and supporting protocols in the DeFi space, allowing them to readily interface with all participants in a decentralized manner.

3.4.1.2 Emergence of Core DeFi Primitives (2017–2018)

This period saw the emergence of foundational DeFi projects that addressed key implementation and adoption challenges in decentralized finance. MakerDAO (Brennecke et al., 2022) launched its decentralized stablecoin, DAI, in 2017. DAI provided a price-stable asset within the volatile cryptocurrency market, which is crucial for facilitating lending, borrowing, and other financial activities. Concurrently, decentralized exchanges (DEXs) (Auer et al., 2024), such as Uniswap (Adams, 2024), began to gain traction. Uniswap's automated market maker (AMM) model has revolutionized peer-to-peer trading by eliminating the need for traditional order books, enabling a truly decentralized, more efficient, and accessible exchange of digital assets.

3.4.1.3 Rapid Expansion and Innovation (2019–2020)

The DeFi landscape experienced significant expansion during this period, driven by the emergence of innovative protocols that expanded the range of financial services offered. For example, Aave (Sun et al., 2022) and Compound (Kao et al., 2020) protocols introduced decentralized lending and borrowing protocols to expand DeFi applications. Although challenging, they enable participants to lend or borrow cryptocurrencies in a permissionless manner. Note that these disruptive protocols introduced the concept of over-collateralized loans, mitigating the risk of default. Furthermore, the rise of yield farming, liquidity mining, and other innovations has incentivized users to provide liquidity to these platforms, fueling DeFi's rapid growth and adoption and driving operational expansion of DEXs (Mirdala, 2024).

3.4.1.4 Mainstream Boom and Continuous Evolution (2021–Present)

In DeFi, Total Value Locked (TVL) measures the value of digital assets staked in smart contracts for lending, borrowing, and trading, reflecting the collateral, liquidity, and utilization of a DeFi system. Because of attracting significant attention from both retail and institutional investors, DeFi experienced explosive growth in TVL during this phase (Auer et al., 2024; Mirdala, 2024). However, this rapid growth also highlighted numerous challenges. In particular, the DeFi space has faced increased regulatory scrutiny as governments worldwide grapple with how to regulate this nascent financial services sector. Security risks, including smart contract vulnerabilities and exploits, have become increasingly prevalent, leading to significant financial losses for users. Scalability also emerged as a major obstacle, as some underlying blockchains struggled to handle increasing transaction volume, leading to high fees and slow processing times. These are priority issues that highlight the challenges in the ongoing evolution of DeFi applications as they strive for greater maturity and stability. Therefore, it is essential for us to highlight the risks and challenges in the Web3 ecosystem more comprehensively.

3.4.2 The Vulnerabilities and Challenges of DeFi in Web3

Decentralized finance (DeFi), while promising, faces a range of significant risks and challenges. To warrant its sustainable growth and mainstream adoption across industries and communities, the following risks and challenges, currently the main concerns, must be addressed quickly and appropriately (Li et al., 2022; Weingärtner et al., 2023; Mirdala, 2024).

3.4.2.1 Smart Contract and Systemic Vulnerabilities

Despite advancements in auditing practices, smart contracts and systemic vulnerabilities continue to pose a critical threat. Code bugs, exploits, and unforeseen events lead to substantial financial losses, as exemplified by high-profile hacks such as the DAO governance attack (Han et al., 2025) and other protocol attacks. As noted by Li et al. (2022), even audited contracts are not immune, underscoring the ongoing need for continually enhanced, robust security audits, formal verification methods, and systematic analysis tools. Inherently, the complexity and frequent revisions of on-chain smart contracts, as well as the rapid pace of DeFi technological advancements, development, and deployment, could largely contribute to this persistent vulnerability.

3.4.2.2 Regulatory Uncertainty

As of 2025, DeFi continues to operate in a largely unregulated landscape, creating uncertainty for both developers and users. The lack of well-defined regulatory frameworks poses compliance challenges and legal ambiguities, potentially hindering innovation and attracting illicit activities (Uzougbo et al., 2024). Jurisdictional variations in regulatory approaches necessitate that DeFi projects would have to navigate a complex and evolving regulatory landscape, further complicating matters. Therefore, the lack of collaboration among regulators, industry, and other stakeholders across the DeFi space could hinder the development of a balanced regulatory framework that supports the growth of DeFi in a safe and responsible manner.

3.4.2.3 Market Risks

The inherent volatility of cryptocurrency markets poses significant risks to DeFi protocols and users. Sudden price swings and liquidity shortages can lead to substantial financial losses and instability, as evidenced by market crashes and cascading liquidations. The interconnectedness of DeFi protocols means that failures in a part of a DeFi deployment can quickly cascade across the Web3 ecosystem, amplifying market risks from multiple perspectives (Weingärtner et al., 2023). Furthermore, despite its decentralized nature, DeFi can exhibit tendencies towards centralization. Concentrated ownership of governance tokens can exert an undue influence on a small group of individuals, thereby undermining the principles of decentralization. Reliance on centralized infrastructure components, such as oracles that provide off-chain data to smart contracts, introduces single points of failure and potential manipulation (Eskandari et al., 2021). Addressing these centralization concerns is critical for maintaining the long-term health and integrity of the DeFi applications.

3.4.2.4 User Experience and Security

The technical complexity of DeFi applications, combined with the need for careful key management, creates barriers to stimulating mainstream adoption. Complex interfaces, coupled with a lack of user-friendly tools, can make it difficult for non-technical users to participate safely. In addition, poor security practices, such as inadequate key management and a lack of education about DeFi, can expose users to theft and fund loss. Thus, improving user experience and simplifying security protocols are crucial for wider adoption (Li et al., 2022; Zheng et al., 2023).

3.4.2.5 Scalability and Performance

Blockchain networks often struggle to handle the growing transaction volume of DeFi applications, leading to high transaction fees, network congestion, and slow processing times. Scalability limitations hinder the efficiency and usability of DeFi protocols and the services they enable. Interoperability issues between different blockchain networks complicate DeFi applications, limiting cross-chain integration and hindering the development of a unified DeFi application (Weingärtner et al., 2023).

3.4.2.6 Final Remarks

Overall, DeFi emblematizes a remarkable innovation in the financial service sector, enabling transparency, accessibility, and efficiency on a decentralized network. However, it also faces substantial risks and challenges, particularly in the context of Web3. Addressing these issues through advancing technology, regulatory clarity, and user education will be critical for the sustainable growth and mainstream adoption of DeFi (Weingärtner et al., 2023).

3.5 Decentralized Applications (DApps)

Decentralized applications (DApps) are a cornerstone of the Web3 ecosystem. As discussed earlier, DApps represent a paradigm shift from traditional, centralized software to applications that operate on decentralized networks. DApps are primarily powered using blockchain technology (Zheng et al., 2023). This transition signifies the adoption of a future digital world, characterized by data-driven, distributed, and participatory co-creation applications. Technically, DApps leverage key Web3 building blocks, such as DAOs, DIDs, and DeFi, to achieve their decentralized nature (Auer et al., 2024). As explored earlier, DAOs enable community governance, DIDs provide user-centric digital identity management, and DeFi integrates decentralized financial services directly into the DApp experience. This well-crafted, collective, collaborative, and integrated architecture eliminates single points of control or failure, fostering community-driven governance and enhancing both resilience and personalized digital user experience in Web3 (Zhou et al., 2023).

From all the discussions above, it is clear that this decentralized architecture empowers users to seamlessly discover, exchange, and manage digital assets with enhanced transparency, security, and trustlessness (Zheng et al., 2023). By removing intermediaries, DApps can facilitate direct peer-to-peer interactions, granting users greater autonomy, privacy, and control over their data and information assets with the support of DAOs and DIDs. As articulated in this book, DApps are built on DAOs, DIDs, and DeFi, becoming open-source, immutable, and capable of integrating native tokens, incentives, or cryptocurrencies for decentralized transactions, community-driven governance, and motivated user participation.

Currently, DApps are revolutionizing various sectors with decentralized solutions. In the DeFi space, platforms such as Uniswap and MakerDAO facilitate peer-to-peer transactions, including trading, lending, and borrowing, without the need for intermediaries. In gaming, DApps such as CryptoKitties, Axie Infinity, and Decentraland use blockchain to verify ownership of in-game assets, allowing players to trade and monetize them (Tempesta, 2024). Social media DApps, such as Steemit, Minds, and Mastodon, offer decentralized platforms for content creation and sharing, often rewarding users with cryptocurrency or other forms of monetized tokens (Akpelu & Godday, 2023). Finally, in supply chain management, DApps such as VeChain and Waltonchain enhance transparency and traceability by recording transactions on their respective blockchains (Hellani et al., 2021).

DApps, while promising a more decentralized and innovative digital landscape, face significant challenges in user experience and development that must be addressed for broader adoption and long-term success. User experience is often hampered by the complexity of managing private keys and interacting with wallets, a significant barrier for nontechnical users. Unpredictable transaction fees ("gas") and slower transaction speeds due to blockchain consensus mechanisms further detract from enhanced usability. In other words, scalability remains a significant bottleneck, which hinders widespread adoption (Arvanitou et al., 2024).

Development challenges are equally crucial. Smart contract security is paramount, as vulnerabilities can lead to devastating exploits. Rigorous testing and audits are essential (Mallare et al., 2023; Popchev & Radeva, 2024). Interoperability, the ability of DApps to interact seamlessly across different blockchains, is another complex technical hurdle. Furthermore, the evolving regulatory landscape creates uncertainty and complexity for developers seeking to comply with varying jurisdictional requirements (Uzougbo et al., 2024). Finally, overcoming the steep learning curve for both users and developers is also a significant obstacle to mainstream adoption.

Addressing these challenges is not only desirable but also essential for the future growth of DApps. Improving user experience through simplified digital "wallet" management, predictable transaction costs, and faster transaction speeds is crucial for expanding the user base. On the development side, robust security practices, scalable blockchain solutions, and standardized interoperability protocols are needed to create a more robust and efficient DApp in the Web3 ecosystem. Furthermore, clearer regulatory frameworks will foster innovation and attract investment. Ultimately, the success of DApps hinges on bridging the gap between the promise of decentralization and the applicability of user adoption and development. Only by tackling these challenges can DApps realize their full potential and usher in a truly decentralized digital future in Web3.

To conclude this section, we can emphasize that DApps will surely embody the core principles of Web3 and foster a more equitable and user-centric digital landscape. They represent a shift from users as passive consumers to active stakeholders in the governance and evolution of the applications they use (Davidson et al., 2018; Qiu, 2024). From finance and gaming to social media and digital identity, DApps drive innovation across a wide range of service sectors. For end-users, they deliver new forms of digital ownership, community participation, and decentralized governance, fundamentally changing how users interact with online services and transforming the future of the Internet (Ray, 2023).

3.6 Conclusion: The Four Pillars of the Web3 Foundational Framework

In this book, we promote and articulate that the Web3 ecosystem can be built upon the assembly and coordination of DAOs, DIDs, DeFi, and DApps, which are defined as the four foundational pillars of Web3. We argue that these pillars collectively and collaboratively ensure data privacy and security, transparency, and value co-creation for all participants (Fig. 3.3):

- DAOs primarily promote and technically enable decentralized, transparent, and inclusive decision-making in organizational governance structures within Web3. Although risks and challenges in DAOs exist and could change, ongoing innovations continue to enhance their functions by addressing these vulnerabilities.
- DIDs deliver an innovative digital identity scheme by providing a secure, privacy-focused, and self-sovereign model, reducing dependence on centralized authorities and fostering trust in online interactions.
- DeFi transforms financial and monetization systems by leveraging decentralized ledgers to offer and deliver equitable value distribution, transparency, and accessibility in financial and monetization transactions.
- DApps help transform users from passive users (i.e., consumers) into active participants (i.e., being both consumers and contributors). DApps facilitate the real-

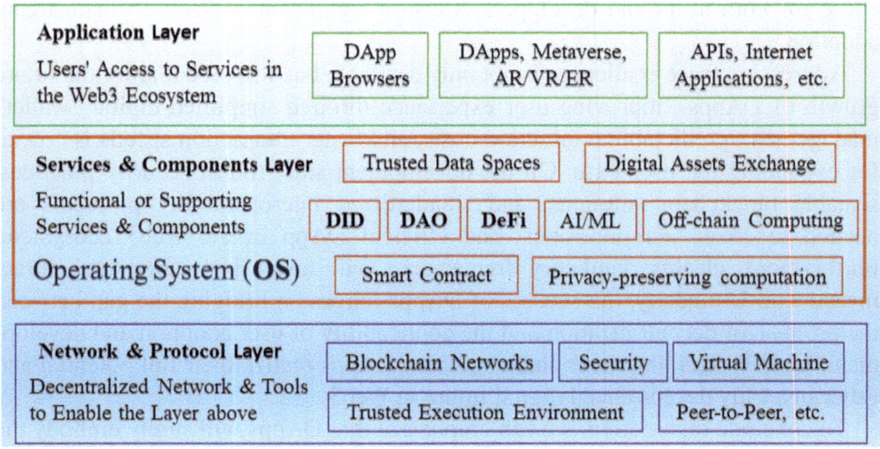

Fig. 3.3 Overview of Web3 foundational pillars by layers

ization of value co-creation, complying with one of the key design and implementation principles of Web3.

Interestingly, the interplay of these four pillars can be understood through the analogy of a modern software application and its operating system (OS). An OS provides a computer application with essential resource control and management services, such as file management, input/out, communication, processes, and security. Computer applications can be developed, deployed, and operated in an effective, efficient, and reliable manner. In the same way, a domain-specific Web3 ecosystem requires its own "OS," which is radically defined by the collective functions of these four pillars:

- DAOs (Governance) create and manage the community that defines operational features, rules, policies, and award mechanisms of the ecosystem.
- DIDs (Identify) provide secure, unique, verifiable, and user-controlled digital identities within the ecosystem.
- DeFi (Finance) services facilitate financial and monetization processes without intermediaries.
- DApps (User Interface) serve as the user-facing layer (i.e., an app) through which you interact, work, or create and share information.

In summary, these pillars form the backbone of the OS of Web3, shaping how the ecosystem functions and, most importantly, how value is co-created by all participants. Just as a computer OS evolves with new tools and capabilities, the Web3 OS will continue to grow with emerging services and components as illustrated in Fig. 3.3, driving the development of future Web3 ecosystems. This foundational framework is crucial for building a more decentralized, efficient, and user-driven digital future.

Table 3.1 A brief comparison of Web2 and Web3 in various aspects

Aspect	Web2 (centralized web)	Web3 (decentralized web)	Continuity
Core concept	Centralized, platform-driven, focused on user content and interactivity	Decentralized, blockchain-based, focused on ownership, autonomy, and peer-to-peer value	Both seek to enhance user engagement and participation
Architecture	Centralized servers and cloud controlled by corporations	Distributed Ledger (Blockchain) and decentralized networks (nodes/communities)	Both require scalable, interconnected network
Data control	Data is owned and controlled by centralized platforms (e.g., Big Tech)	Users own and manage their data via cryptography and decentralized storage	Both rely on large-scale data for operations
Trust mechanism	Relies on institutional trust (intermediaries, companies)	Relies on trustless systems (cryptography, smart contracts, consensus)	Both need mechanisms for trust and accountability
Monetization model	Platform-centric: Revenue from ads, subscriptions, and data exploitation	Token-based economy: Value earned via tokens, NFTs, and DeFi	Both support digital commerce and service monetization
User role	Consumers and content creators within closed platforms	Active Participants, stakeholders, and co-owners (DAOs)	Both empower user contribution and creativity
Identity management	Centralized Identity; multiple logins managed by platforms	Self-sovereign identity using Decentralized Identifiers (DIDs)	Both aim to authenticate and personalize the experience
Data portability	Limited; data is often locked into specific platforms	High; users can move and reuse data freely across platforms	Both depend on data interoperability.
Technology stack	Web browsers, APIs, cloud computing, social media	Blockchain, smart contracts, AI/ML, privacy computing	Both are built upon core Internet and networking technologies
Governance	Managed by corporations or centralized authorities	Community-driven via Decentralized Autonomous Organizations (DAOs)	Both involve structured decision-making
Intelligence	Search engines, recommendation systems, Big Data analytics	AI/ML integrated with blockchain for intelligent automation	Both use data intelligence to improve the user experience
Economic model	Platform Capitalism: Profit concentrated in major tech firms	Token Economy: Value distributed among users and stakeholders	Both create new digital markets and business models
Vision	Connectivity, participation, and collaboration	Decentralization, ownership, and empowerment	Both aim for a more interactive and human-centric Internet

Given that Web3 is still emerging, we know that all the explorations so far, starting with Chap. 1, will surely require continuous updates as time goes on. However, the fundamental concepts and technologies are most likely to remain for the foreseeable future. Before we jump into the discussion of the future Web3 ecosystems, a clear, foundational understanding of Web3 is essential. To help readers grasp the concepts and principles of Web3 and clearly understand the differences between the Internet migrating from Web2 to Web3, a brief comparison between Web2 and Web3 is presented in Table 3.1.

From the above table, we know that Web2 is essentially the interactive and social web we are currently using, defined by platforms that harness user-generated content to create value, which is, however, centralized in the hands of a few powerful companies. We must also understand that Web3 is the decentralized, user-centric web of the future, leveraging blockchain technology and privacy-preserving computation to return ownership and control to users, creating an open, trustless, and permissionless Internet.

The next two chapters present case studies demonstrating how real-world Web3 ecosystems, built on these pillars, are designed, implemented, and operated to enable collaborative value co-creation.

References

Adams, A. (2024). Layer 2 be or Layer not 2 be: Scaling on Uniswap v3. *arXiv preprint arXiv:2403.09494*.

Akpelu, F. E., & Godday, E. I. (2023). Enhancing administrative heads of public tertiary institutions in rivers state for effective information dissemination through decentralized social media networks. *Contemporary Accounting and Management Research Journal, 7*(3), 25–42.

Anwar, H. (2018). *Web 3.0 Blockchain Technology Stack: The Comprehensive Guide*. Retrieved Jan. 10, 2025, from https://101blockchains.com/web-3-0-blockchain-technology-stack/

Arvanitou, E. M., Gagoutis, D., Ampatzoglou, A., Mittas, N., Deligiannis, I., & Chatzigeorgiou, A. (2024). What does matter in the success of a decentralized application? From idea to development. *Information and Software Technology, 169*, 107414.

Auer, R., Haslhofer, B., Kitzler, S., Saggese, P., & Victor, F. (2024). The technology of decentralized finance (DeFi). *Digital Finance, 6*(1), 55–95.

Avellaneda, O., Bachmann, A., Barbir, A., Brenan, J., Dingle, P., Duffy, K. H., et al. (2019). Decentralized identity: Where did it come from and where is it going? *IEEE Communications Standards Magazine, 3*(4), 10–13.

Bellavitis, C., Fisch, C., & Momtaz, P. P. (2023). The rise of decentralized autonomous organizations (DAOs): A first empirical glimpse. *Venture Capital, 25*(2), 187–203.

Brennecke, M., Guggenberger, T., Schellinger, B., & Urbach, N. (2022). The de- central bank in decentralized finance: A case study of MakerDAO. *Proceedings of the 55th Hawaii International Conference on System Sciences*, 6073–6082.

BTH. (2024). *Web3 and the future of work: Decentralized autonomous organizations (DAOs)*. Retrieved on Feb. 1, 2025, from https://beyondthehype.terrencegatsby.com/web3/web3-and-the-future-of-work- decentralized-autonomous-organizations-daos/

Buterin, V. (2014). A next-generation smart contract and decentralized application platform. *Ethereum White Paper, 3*(37), 2–1.

Davidson, P., De Filippi, P., & Potts, J. (2018). The economics of Blockchain. *Journal of Technology Transfer, 43*(5), 1246–1267.

Ding, W., Hou, J., Li, J., Guo, C., Qin, J., Kozma, R., & Wang, F. Y. (2022). DeSci based on Web3 and DAO: A comprehensive overview and reference model. *IEEE Transactions on Computational Social Systems, 9*(5), 1563–1573.

Dock (2024). Web3 Identity: Beginner's Guide 2024. Retrieved on Feb. 8, 2025, from https://www.dock.io/post/web3-identity#what-is-web3-identity-and-why-do-we-need-it

Eskandari, S., Salehi, M., Gu, W. C., & Clark, J. (2021). Sok: Oracles from the ground truth to market manipulation. In *Proceedings of the 3rd ACM Conference on Advances in Financial Technologies* (pp. 127–141).

Ghosh, A., Hassija, V., Chamola, V., & El Saddik, A. (2024). A survey on decentralized metaverse using Blockchain and Web 3.0 technologies, applications, and more. *IEEE Access, 12*, 146915–146948.

Han, J., Lee, J., & Li, T. (2025). A review of DAO governance: Recent literature and emerging trends. *Journal of Corporate Finance, 102734.*

Hellani, H., Sliman, L., Samhat, A. E., & Exposito, E. (2021). On blockchain integration with supply chain: Overview on data transparency. *Logistics, 5*(3), 46.

Huang, Y., Bian, Y., Li, R., Zhao, J. L., & Shi, P. (2019). Smart contract security: A software lifecycle perspective. *IEEE Access, 7*, 150184–150202.

Kao, H. T., Chitra, T., Chiang, R., & Morrow, J. (2020, May) An analysis of the market risk to participants in the compound protocol. In *Third international symposium on foundations and applications of blockchains* (Vol. 362).

Kim, J. (2021). Regulation of decentralized systems: A study of Uniswap. *Harvard Journal of Law & Technology, 35*, 335.

Kozhan, R., & Viswanath-Natraj, G. (2022). Fundamentals of the MakerDAO governance token. In *3rd International Conference on Blockchain Economics, Security and Protocols (Tokenomics 2021).* Schloss-Dagstuhl-Leibniz Zentrum für Informatik.

Krause, D. (2024). *The Rise of Web3: Opportunities and Challenges.* Retrieved on Feb. 1, 2025, from https://ssrn.com/abstract=4895353

Li, W., Bu, J., Li, X., Peng, H., Niu, Y., & Zhang, Y. (2022). A survey of DeFi security: Challenges and opportunities. *Journal of King Saud University- Computer and Information Sciences, 34*(10), 10378–10404.

Mallare, D., Dani, N., & Bonaci, T. (2023). Decentralized identifiers. *IEEE Potentials, 42*(6), 18–24.

Mazzocca, C., Acar, A., Uluagac, S., Montanari, R., Bellavista, P., & Conti, M. (2024). A survey on decentralized identifiers and verifiable credentials. arXiv preprint arXiv:2402.02455.

Mirdala, R. (2024). Revolutionizing finance: Decentralized finance as a disruptive challenge to traditional finance. *Theoretical and Practical Research in Economic Fields, 15*(3), 517–538.

Nakamoto, S. (2008). Bitcoin: A peer-to-peer electronic cash system. *Satoshi Nakamoto.*

Popchev, I., & Radeva, I. (2024). Decentralized application (dApp) development and implementation. *Cybernetics and Information Technologies, 24*(2), 122–141.

Qin, R., Ding, W., Li, J., Guan, S., Wang, G., Ren, Y., & Qu, Z. (2023). Web3- based decentralized autonomous organizations and operations: Architectures, models, and mechanisms. *IEEE Transactions on Systems, Man, and Cybernetics: Systems, 53*(4), 2073–2082.

Qiu, R. G. (2014). *Service science: The foundations of service engineering and management.* Wiley.

Qiu, R. G. (2024). Web3 and service science: Perfectly aligned to maximize the values of all stakeholders in the digital economy. *Digital Transformation and Society, 3*(4), 337–339.

Ray, P. P. (2023). Web3: A comprehensive review on background, technologies, applications, zero-trust architectures, challenges and future directions. *Internet of Things and Cyber-Physical Systems, 3*, 213–248.

Rikken, O., Janssen, M., & Kwee, Z. (2023). The ins and outs of decentralized autonomous organizations (DAOs) unraveling the definitions, characteristics, and emerging developments of DAOs. *Blockchain: Research and Applications, 4*(3), 100143.

SAFIT. (2025). *ChainWeaver: Weaving Chains into a Network and Sharing Connectivity.* Retrieved Jan. 10, 2025, https://docs.chainweaver.org.cn/index.html

Schär, F. (2021). Decentralized finance: On blockchain- and smart contract-based financial markets. *Federal Reserve Bank of St Louis Review, Second Quarter, 103*(2), 153–174.

Sporny, M., Guy, A., Sabadello, M., & Reed, D. (2022). Decentralized identifiers (DIDs) v1.0, Core architecture, data model, and representations. *World Wide Web Consortium (W3C).* Retrieved on Feb. 8, 2025, https://www.w3.org/TR/did-1.0/

Sun, X., Stasinakis, C., & Sermpinis, G. (2022). Liquidity risks in lending protocols: Evidence from aave protocol. arXiv preprint *arXiv:2206.11973.*

Surve, T., & Khandelwal, R. (2023). The development of decentralized governance models for web 3 ecosystems. In *Concepts, technologies, challenges, and the future of Web 3* (pp. 91–107). IGI Global.

Tang, C., Cai, Q., Dong, C., Wang, Q., & Chen, S. (2025). Decentralised autonomous organizations (DAOs): An exploratory survey. *Distributed Ledger Technologies: Research and Practice.*

Tempesta, S. (2024). *Application architecture patterns for Web 3.0: Design patterns and use cases for modern and secure Web3 applications*. CRC Press.

Uzougbo, N. S., Ikegwu, C. G., & Adewusi, A. O. (2024). Regulatory frameworks for decentralized finance (DEFI): Challenges and opportunities. *GSC Advanced Research and Reviews, 19*(02), 116–129.

Vasileva, H. (2025). Uniswap becomes the leading DAO, large treasuries raise the issue of governance attacks in 2025, Retrieved on Feb. 1, 2025, from https://www.cryptopolitan.com/uniswap-dao-governance-attacks-2025/

Wang, S., Ding, W., Li, J., Yuan, Y., Ouyang, L., & Wang, F. Y. (2019). Decentralized autonomous organizations: Concept, model, and applications. *IEEE Transactions on Computational Social Systems, 6*(5), 870–878.

Weingärtner, T., Fasser, F., Sá, R., da Costa, P., & Farkas, W. (2023). Deciphering DeFi: A comprehensive analysis and visualization of risks in decentralized finance. *Journal of Risk and Financial Management, 16*(10), 454.

Wright, A. (2021). The rise of decentralized autonomous organizations: Opportunities and challenges. *Stanford Journal of Blockchain Law & Policy, 4*, 1.

Yaga, D., Mell, P., Roby, N., & Scarfone, K. (2019). *Blockchain technology overview*. arXiv preprint arXiv:1906.11078.

Zheng, P., Jiang, Z., Wu, J., & Zheng, Z. (2023). Blockchain-based decentralized application: A survey. *IEEE Open Journal of the Computer Society, 4*, 121–133.

Zhou, L., Xiong, X., Ernstberger, J., Chaliasos, S., Wang, Z., Wang, Y., et al. (2023). Sok: Decentralized finance (defi) attacks. In *2023 IEEE Symposium on Security and Privacy (SP)* (pp. 2444–2461). IEEE.

Value Co-creation Case Study: The Smart Car Ecosystem

4

Abstract

In this chapter, we present a case study of the emerging smart car ecosystem. While it promises safer, more efficient mobility, its current deployment is hindered by centralized data governance and fragmented infrastructure. We examine how Web3 technologies, particularly Decentralized Autonomous Organizations (DAOs), Decentralized Identifiers (DIDs), decentralized finance (DeFi), and decentralized applications (DApps), can address these challenges by shifting from a siloed corporate model to a decentralized, value co-creation framework. More importantly, using the conceptual *AutoQDrive* mobility platform, built on the pillars of Web3, we demonstrate three transformative use cases of value co-creation in the smart car ecosystem. First, asset digitization tokenizes underutilized resources (such as vehicles and infrastructure) into NFTs, which are managed via DeFi. As a result, resource utilization is optimized, creating an innovative revenue stream. Secondly, community-driven governance and tokenized incentives (*QCoins*) motivate users to share their mobility data, transforming it into a collective-value asset that supports more smart services, including intelligent traffic management, personalized navigation, and predictive maintenance. Thirdly, participatory decision-making fosters inclusivity by enabling stakeholders to collaboratively plan and design future critical infrastructure, such as smart parking facilities, in the smart car ecosystem.

Keywords

Smart car ecosystem · Tokenized incentives · Asset digitization · Data value · Resource utilization · Participatory decision-making

R. Qiu, *Value Co-Creation in Web 3.0*, Advances in Data Analytics, AI, and Smart Systems, https://doi.org/10.1007/978-3-032-12911-6_4

4.1 Introduction to a Smart Car Ecosystem

The Motorwagen, powered by an internal combustion engine, is widely regarded as the first modern automobile, designed and built by Karl Benz in 1885. A transformative era in the automotive industry followed the introduction of mass-production techniques pioneered by Henry Ford in 1908. By launching the Model T, Ford introduced and emphasized simplicity, basic functionality, and large-scale automobile production, with the purpose of serving the general population. The adoption of an innovative assembly line indeed drastically improved production efficiency and vehicle quality, while reducing manufacturing costs. The resulting affordability of the Model T sparked a profound societal and economic transformation in the United States in the early twentieth century (Lonergan, 2021). Today, the assembly line production model has been widely adopted, substantially improved in terms of automation, and innovatively applied not only in the automotive industry but also across many other industrial sectors, making automobile ownership more accessible to the average person and reshaping manufacturing, transportation, and daily life worldwide.

During the 1970s, the integration of electronics into automobiles marked a significant turning point in the automotive industry's evolution. Innovations such as electronic fuel injection and antilock braking systems signified the beginning of a new era, in which electronic systems became increasingly essential for vehicle performance and driver safety (Lonergan, 2021). With the accelerated globalization and advancements in electronic technology, vehicle engineering and technology driven by electronics have become increasingly pervasive in vehicle designs, production, and usage, highlighting the growing reliance on electronics to enhance efficiency, control, quality, and safety, as well as cost reduction due to the global cooperation in supply chains. Over the years, we have witnessed the innovations and advancements in electronics that have laid the solid foundation for the modern, technology-driven automobile.

Growing concerns about global sustainability, coupled with rapid advancements in digital technologies in the early twenty-first century, have driven a significant transformation in the automotive industry. This shift, both digital and environmental, is exemplified by the emergence of hybrid vehicles, such as the Toyota Prius, and the rise of fully electric models, like Tesla's 2008 Roadster. These innovations marked a pivotal move toward cleaner, smarter mobility solutions, reflecting the industry's evolving priorities in energy efficiency, emissions reduction, and intelligent vehicle design and production (Romeu Lezama, 2016).

An ecosystem is an interconnected network of components (living and/or nonliving) that interact with one another and their environments to function as a cohesive whole. The term originates in biology but is now widely used metaphorically across fields such as technology, business, and social systems. In the context of intelligent transportation systems, this concept is particularly relevant. In addition to the advancements in materials science, electronics, and vehicle engineering, the confluence of rapid developments in networking, communications, computing, pervasive sensing, and data analytics has been a significant catalyst for the modern automobile industry. This leap in progress enables the seamless integration of diverse components, including technologies, organizations, users, regulations, and transportation infrastructure (Dong & Qiu, 2004).

Specifically, within this early proposed transportation ecosystem, vehicles, software, and users can collaborate to co-create value, solve problems, and enable innovation. For example, this collaboration can manifest in systems that provide point-of-need information services on highways. The introduction of ubiquitous mobile and pervasive sensing technologies has enabled the development of a network that provides drivers and passengers with immediate access to vital information, such as hospital locations in unfamiliar areas, or with immediate assistance and guidance at the point of need during emergencies such as hazardous material spills in transit (Dong & Qiu, 2004).

Based on research by Dong and Qiu (2004), this early ecosystem prototype functions through the coordinated interplay of various components:

- Users (e.g., drivers, passengers, emergency responders) who need and utilize the information services
- Technologies such as radio frequency identification (RFID), agent technologies, communication infrastructure (including the Internet and wireless networks), and computing devices (such as smartphones, navigation systems, and edge computing devices)
- Service providers that offer information on travel, traffic, and local services, as well as emergency response and product details
- The transportation infrastructure, including vehicles, highways, roads, sensors, and traffic controllers, i.e., the physical environment in which these interactions in the ecosystem occur

The efficiency and effectiveness of this early ecosystem prototype hinge on the seamless flow of data and communication, which ensures that relevant knowledge can be extracted and delivered correctly in real time, precisely when and where it is needed. Currently, the emerging smart car ecosystem, anchored by four foundational pillars (i.e., connectivity, autonomy, software-centricity, and ecosystem integration), represents a transformative frontier in the ongoing digital mobility revolution.

Firstly, at its core, connectivity is powered by technologies such as the Internet of Things (IoT), 5G, and Vehicle-to-Everything (V2X) technologies, enabling vehicles to interact not only with each other but also with roadside infrastructure, pedestrians, and cloud platforms (Khan et al., 2021). This continuously increased resolution and data quality in interconnectivity enable real-time data exchange, predictive diagnostics, and adaptive route optimization.

Secondly, autonomy leverages artificial intelligence (AI), sensor fusion, and advanced driver-assistance systems (ADAS) to deliver increasingly self-sufficient driving experiences. These systems operate through intelligent perception and decision-making, supported by massive real-time data, algorithms, machine learning, and edge computing, minimizing reliance on human input while responding in milliseconds to dynamic road conditions (Zhao et al., 2020).

Thirdly, recently emerged software-centric models have revolutionized the redefinition of vehicle operational architecture. Through over-the-air updates and modular firmware, automakers now deploy features, security patches, preventive maintenance, and performance enhancements without requiring physical service appointments. This continuous software update integrates vehicles into digital service platforms, creating user-centric environments that are deeply customized to meet the needs of end-users, thereby enhancing convenience and safety (Ghosh et al., 2022).

Finally, smart cars are no longer standalone machines. Each smart vehicle is an embedded node in a larger socio-technical service ecosystem. Building on early prototypes, such as intelligent highway point-of-need systems, vehicles are envisioned as active personalized participants in the real-time, context-aware service ecosystem (Dong & Qiu, 2004). Today, informed in real time by sensor data and user interactions with the physical environment (e.g., RFID-tagged infrastructure, roadside sensors), this ecosystem provides dynamic information and responsive support tailored to moment-specific needs for all stakeholders within the ecosystem.

By integrating with urban systems, such as energy grids, public transportation, and emergency services, the smart car ecosystem becomes a key enabler of sustainable, smart cities. It facilitates not only safer and more efficient mobility but also new paradigms in human-centered service system design (Qiu, 2014), where physical and digital touchpoints converge to deliver smart *deepservice*, a deeply situated, adaptive, responsive, intelligent, and enjoyable service experience.

To set the stage for a deeper exploration of value co-creation in the smart car ecosystem within the Web3 landscape, the remainder of this section provides a concise yet insightful examination of the smart car ecosystem. It begins with an overview of its core components and underlying technologies, followed by a discussion of the key benefits and challenges that drive its ongoing evolution.

4.1.1 Key Components and Underlying Technologies of the Smart Car Ecosystem

The rapid evolution of the smart car ecosystem, a dynamic socio-technical system since the turn of the millennium, is largely driven by significant technological advancements across its interconnected core components, including vehicle connectivity, automated driving, transportation infrastructure, and big data computing. In fact, it is the powerful combination of these integrated components that drives the emergence of a vibrant and increasingly safe mobility ecosystem.

Connectivity, a key component, plays a crucial role in enhancing the intelligence of modern vehicles and improving their user experience. Specifically, equipped with a comprehensive suite of network-based sensors, GPS, and robust cellular/Wi-Fi connectivity, most recently built vehicles can exchange real-time data across the mobility landscape. For example, we have observed that Vehicle-to-Vehicle (V2V) communication is well-suited for cooperative driving, while Vehicle-to-Infrastructure (V2I) interaction with smart roads and traffic systems is utilized for enhanced road traffic control and management. Additionally, V2X technology is widely used to enhance connectivity for pedestrians and road environments (Souri et al., 2024). Notably, the substantially improved interconnectedness has already significantly contributed to enhanced safety, optimized traffic flow, and the delivery of real-time information in many metropolitan areas worldwide.

Since the turn of this millennium, autonomous driving has become another key component in advancing the smart car ecosystem. These capabilities are steadily advancing. Smart cars utilize a sophisticated array of sensors, including laser-based Light Detection and Ranging (LiDAR), cameras, and network-based sensor devices for comprehensive environmental perception. All collected sensor data is processed by powerful AI and advanced algorithms, enabling vehicles to navigate, make critical decisions, and execute maneuvers with increasing autonomy, gradually reducing the need for human intervention. Current AI-powered self-driving features, such as Tesla Autopilot and Waymo, represent significant leaps toward fully autonomous driving in the foreseeable future (Shanmugam et al., 2024).

The unceasingly improved smart transportation infrastructure is the intelligent backbone of today's smart car ecosystem. For instance, beyond traditional traffic control signals, the ecosystem includes sensors and data collection devices that monitor road conditions, the status of smart charging stations, and mass flow. Many initiatives also focus on developing smart cities by integrating smart city mobility and urban public service networks, facilitating the development of adaptive urban systems (Bittencourt et al., 2024). As a result, more intelligent transportation service systems can be designed and developed to assist drivers and passengers at the point of demand (Dong & Qiu, 2004). Notably, connecting vehicles seamlessly to this network-based, sensor-enabled infrastructure fosters real-time data exchange and analysis, significantly enhancing vehicle performance, public safety, and the ecosystem's overall efficiency (Butler et al., 2020).

As the data-driven paradigm becomes the norm, cloud and big data platforms have gradually grown to become the central nervous system of the smart car ecosystem. These powerful platforms can now process and analyze vast volumes of data generated by connected vehicles in real time, enabling intelligent decision-making across multiple dimensions (Alexakis et al., 2022). For example, these platforms provide advanced analytics for traffic optimization, early detection of mechanical issues through predictive maintenance, and the delivery of highly personalized user experiences (Abirami et al., 2024). Moreover, as mentioned earlier, over-the-air software updates that can be streamed seamlessly from the cloud are rapidly becoming standard practice, enabling continuous feature upgrades, performance improvements, and critical security patches at the point of demand. It is worth mentioning that, with the interconnected nature of this socio-technical system, robust protection against hacking and data breaches is a critical imperative for ensuring the safety of people and the sustainability of the ecosystem.

The goal of the smart car ecosystem is to create a safer, cleaner, and more efficient future for transportation by seamlessly integrating advanced technology, constant connectivity, and intelligent infrastructure, thereby further expanding the ecosystem while stimulating future economic growth and technological innovation (Alam et al., 2024). To the public, when articulated in simple terms, the ecosystem must focus on delivering a positive and enriched user mobility experience. This includes mobile applications that enable remote vehicle management, preventive maintenance, intelligent route optimization, and access to flexible ride-sharing options (e.g., carpooling). When examining the interior of the vehicle, advanced infotainment systems, responsive voice assistants, and deeply personalized settings will elevate the driving and riding experience from a mere task to a highly engaging, comfortable, and profoundly user-centric experience. Ultimately, value co-creation innovations can be optimally materialized to facilitate the development of a promising future for mankind.

4.1.2 Benefits and Challenges of the Smart Car Ecosystem

The smart car ecosystem is revolutionizing transportation by converging vehicles, infrastructure, and data into a cohesive, intelligent network, yielding significant benefits across multiple dimensions. Evidently, safety is significantly enhanced by advanced collision-avoidance systems, the error-reducing potential of autonomous driving technologies, and real-time hazard and safety alerts disseminated via V2X communication. Mobility efficiency is significantly improved through intelligent routing, adaptive traffic signals, and proactive congestion management, resulting in tangible reductions in commute times, fuel consumption, and harmful emissions. Sustainability is further fostered by ADAS, which promotes energy-efficient driving, the accelerated adoption of electric vehicles (EVs), and optimized energy management. Those advances collectively contribute to substantial decreases in greenhouse gas emissions and improved air quality (Lonergan, 2021; Alam et al., 2024).

As briefly explored earlier, the smart car ecosystem offers individuals a new level of convenience. This is delivered through advanced features, including autonomous parking, seamless over-the-air software updates, and intuitive remote control via a mobile application. Economic advantages materialize through lower operating costs associated with EVs, fairer usage-based insurance models, and cost-effective shared mobility options like ridesharing, easing the financial burden of personal vehicle ownership. Furthermore, data-driven insights unlock proactive predictive maintenance, highly personalized in-vehicle experiences, and more informed urban mobility planning strategies.

Beyond individual gains, the smart car ecosystem promotes greater social equity by offering enhanced mobility solutions for underserved populations, including the elderly and individuals with disabilities, through autonomous transportation and affordable ridesharing services. Public health outcomes are positively impacted by the reduction in air pollution and the alleviation of commuting stress. Underpinning these benefits are robust cybersecurity protocols ensuring the protection of personal data, while continuous innovation in AI, connectivity, and infrastructure fuels the sustainable and long-term expansion of this transformative ecosystem (Alam et al., 2024). Integrated seamlessly with smart city public transit systems, the smart car ecosystem cultivates resilient and user-centric mobility networks, ultimately delivering safer, cleaner, more efficient, equitable, and convenient transportation, thereby reshaping the fabric of urban living.

While the smart car ecosystem offers significant promise, its development is not without its challenges. Key obstacles include navigating complex regulatory landscapes, mitigating cybersecurity risks, ensuring infrastructure readiness, and fostering effective cross-industry collaboration. To unlock the car ecosystem's transformative potential, these issues must be addressed proactively, realizing the promising benefits of enhanced user experience, sustainable transportation development, and improved urban mobility.

4.1.2.1 Regulatory Compliance: Navigating the Evolving Legal Landscape

The rapid advancement of smart car technologies necessitates regulatory frameworks that are not only adaptive and future-oriented but also capable of striking a balance between innovation, public safety, and legal accountability (Van Den Boom, 2022). Critical areas requiring immediate and sustained regulatory attention include the development of harmonized global standards for autonomous driving, robust mechanisms for data ownership and data-use governance, and the clear attribution of liability in the event of autonomous vehicle incidents (Martinho et al., 2021). Additionally, the integration of real-time safety protocols, enabled by V2X communication, necessitates legal provisions that support dynamic, context-aware responses to road conditions. Inconsistent or outdated regulatory approaches across jurisdictions risk undermining deployment, creating legal ambiguity, and delaying the transformative impact of the ecosystem. Future regulatory strategies must emphasize interoperability, scalability, and cross-border coherence to effectively support the continued evolution of smart mobility ecosystems.

4.1.2.2 Security and Data Privacy: Fortifying Connected Mobility Against Evolving Threats

As smart cars become increasingly interconnected with digital infrastructure, they emerge as high-value targets for sophisticated cyberattacks. Protecting these vehicles demands the deployment of robust, multilayered cybersecurity architectures that encompass hardware, software, and communication protocols to safeguard against system breaches, unauthorized access, and malicious manipulation that could endanger both passengers and broader transportation networks (Alam, 2024). The complexity of securing in-vehicle networks, combined with wireless interfaces like V2X and IoT, further intensifies the risk landscape.

Simultaneously, the collection, processing, and sharing of vast volumes of granular user personal data, ranging from driving behaviors to location patterns, raises profound privacy concerns. Ensuring ethical and secure data management necessitates the implementation of transparent, user-centric governance frameworks that comply with stringent global privacy regulations (e.g., GDPR, CCPA), while also leveraging advanced privacy-enhancing technologies such as anonymization, pseudonymization, and differential privacy (Razi et al., 2025). Maintaining user trust through demonstrable privacy protection and system resilience is not only a regulatory necessity but also a critical enabler for widespread adoption of smart mobility solutions over time.

4.1.2.3 Infrastructure Readiness: Building the Foundation for Intelligent Mobility

The successful deployment of the smart car ecosystem is fundamentally dependent on the readiness, modernization, and sustainability of the supporting infrastructure (Butler et al., 2020). To fully enable connected vehicles, significant, coordinated investment in intelligent transportation systems is required. These include sensor-integrated roadways that deliver real-time environmental and traffic insights, intelligent traffic light systems that optimize vehicular flow based on dynamic conditions, and digital signage that communicates directly with vehicles via V2X protocols.

High-reliability, low-latency connectivity, enabled by evolving 5G Advanced and future 6G networks, is crucial for supporting seamless data transmission, mission-critical safety functions, and real-time decision-making in autonomous driving scenarios (Damaj et al., 2022). Moreover, the proliferation of fast-charging and wireless charging stations is essential to support widespread EV adoption, reduce range anxiety, and integrate sustainable mobility into people's daily urban mobility planning.

Additionally, the development of secure, scalable cloud platforms with distributed edge computing capabilities is necessary to manage the massive volumes of data generated by smart vehicles and infrastructure. These platforms enable decentralized processing, providing responsive services such as predictive traffic control, real-time navigation updates, and ADAS, while maintaining data security and minimizing latency to deliver promised and improved user experience.

4.1.2.4 Cross-Industry Collaboration: Orchestrating a Unified Ecosystem

The multifaceted nature of the smart mobility ecosystem necessitates an unprecedented level of cross-industry collaboration, transcending traditionally siloed domains and embracing the principles of Web3 for enhanced data-driven value co-creation. This involves the coordinated efforts of automotive manufacturers, software and AI developers, sensor and hardware providers, telecommunications operators, energy suppliers, urban planners, regulatory authorities, and, crucially, end-users within a more participatory framework. Seamless interoperability and system-wide integration will increasingly rely on the establishment of decentralized and transparent, shared technical standards, open and permissionless innovation platforms leveraging blockchain technologies, and collectively governed infrastructure planning and strategies (Alam et al., 2024; Singh et al., 2025).

Web3 principles can naturally enhance this collaboration by enabling novel value co-creation models. For instance, decentralized data marketplaces could facilitate secure and transparent data sharing among stakeholders, with value distributed based on contributions (Qiu, 2024; SAFIT, 2025). Tokenized incentives could reward participation in open innovation challenges and the development of interoperable solutions. Smart contracts can automate agreements and ensure transparent revenue sharing among all diverse and participating contributors.

Without a unified and collaboratively developed approach, the ecosystem risks fragmentation, as evidenced by incompatible systems, disjointed data flows, and delayed deployment of critical services. Strong, trust-based partnerships and inclusive, decentralized governance frameworks, leveraged through DAOs and DIDs, are thus essential for navigating the challenges of technological integration, securing and auditable data sharing, and transparent liability management.

By fostering collaborative ecosystems rooted in transparency, interoperability, co-innovation, and the value co-creation potential of Web3 (Qiu, 2024), stakeholders can collectively accelerate the transition to a resilient, sustainable, and truly user-centric future of mobility. In this paradigm, smart vehicles and infrastructure function as an integrated whole, with value generated and shared across a diverse network of contributors, ultimately benefiting all users and fostering a more equitable and innovative smart mobility landscape, as promised, in the Web3 era.

4.2 Web3-Driven Value Co-creation in the Smart Car Ecosystem

The successful standardization and development of the smart car ecosystem is inextricably interwoven with all stakeholders, including governments, automakers, research and development institutions, communication and network service providers, and road operators. In other words, the development of common frameworks and technical standards to achieve a truly shared vision largely depends on the synergistic coordination and collaboration among these participants and stakeholders. For example, the evolution of 5G/6G networks and advanced V2X communication protocols is crucial to building low-latency, secure, and interoperable mobility infrastructures. If the necessary collaborative foundation does not exist, networking innovations and the developed products for the proposed ecosystem will become fragmented, which would hinder their reliability, limit their scalability, and erode public trust.

Notably, Web3 technologies provide a transformative foundation for driving value co-creation innovation, which is essential for developing the future smart car ecosystem. More specifically, by leveraging the four foundational pillars discussed in Chap. 3, a Web3-empowered smart car ecosystem can enable more inclusive, secure, and participatory forms of governance and service delivery across diverse stakeholders. Once again, as highlighted by Qiu (2024), the architectural philosophy of Web3 aligns seamlessly with the tenets of Service Science, which emphasizes human-centricity and the co-creation of systemic value within service ecosystems. Therefore, this alignment positions Web3 as a catalyst for transitioning the smart mobility landscape from a traditionally centralized and siloed model to a more collaborative, equitable, and value-driven co-creation paradigm, empowering not only industry and regulatory bodies but also end-users to actively participate in shaping and benefiting from the smart car ecosystem.

This paradigm shift supports more than technological advancement; it reorients the development of intelligent mobility toward a human-centered, AI-augmented, and data-responsible future. It paves the way for what Qiu (2024) refers to as smart *deepservice*, highly personalized, ethically engineered, and contextually relevant services that enrich and safeguard people's daily lives. With this spirit, the following remaining section explores three potential Web3-driven use cases that exemplify how value co-creation can be realized in the smart car ecosystem, marking a significant step toward a more intelligent and participatory future of city mobility in the Web3 era.

4.2.1 Asset Digitization, Sharing, and Transfer to Maximize Value Co-creation

To fully realize resource sharing and maximize asset utilization within the smart car ecosystem, underutilized physical assets, such as vehicles or public mobility infrastructure, can be monetized to create sharing value. Monetization through DeFi involves representing these tangible assets as unique digital tokens, often non-fungible tokens (NFTs), which securely encode ownership, pricing information, and usage rights on a blockchain within the ecosystem. For example, a privately owned vehicle that is frequently idle can be monetized through tokenization, with its digital representation enabling traceable, verifiable, and transferable rights on the network. These tokens can then be exchanged or leased within the smart car ecosystem, fostering value co-creation and a more efficient and decentralized allocation of capital and infrastructure in the community.

Consider a city known as "QCity," which has implemented a Web3-empowered mobility platform called *AutoQDrive*, leveraging DAO, DID, DeFi, and DApp, to manage and optimize its smart car ecosystem. This integrated platform connects data from a wide range of sources, such as connected vehicles, roadway and roadside sensors, adaptive traffic control systems, smart parking infrastructure, and public transit networks, to deliver intelligent mobility services while actively incentivizing user participation and engagement.

One of the transformative capabilities enabled by *AutoQDrive* is the tokenization of underutilized assets, including privately owned vehicles and publicly shared infrastructure. This mechanism aligns with Web3's core tenets of user empowerment, transparency, and decentralized control over digital assets. In a practical scenario (Fig. 4.1), an individual or business entity in QCity could acquire an NFT representing an underutilized vehicle. The NFT encapsulates the vehicle's ownership and usage rights, which are securely managed and transparently recorded on the blockchain.

The NFT holder could then leverage this digital asset to establish a car rental microservice by integrating with a DApp within *AutoQDrive*, such as a blockchain-based rental management platform (Shrivastava et al., 2024). This integration

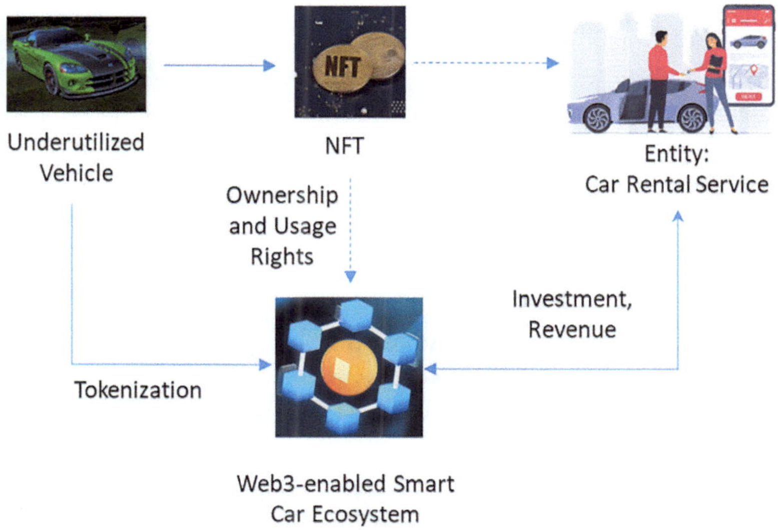

Fig. 4.1 Tokenization for resource optimization and value co-creation

enables automated features, including smart booking, payments, identity verification, and maintenance tracking, through the use of smart contracts. In return for their contribution, the asset holder earns a share of the revenue generated by the service, establishing a value co-creation model in which multiple stakeholders—asset owners, service providers, and end-users—mutually benefit.

Furthermore, the use of smart contracts ensures frictionless and secure asset transfers among ecosystem participants, fostering inclusive, adaptive, and demand-responsive innovation in mobility services. This decentralized, token-driven framework not only optimizes resource utilization but also redefines how value is co-created and distributed in the urban mobility landscape.

4.2.2 Data-Driven Service Improvement and Ownership Incentives

The fast advances of smart car technologies have led to an explosion in data generation, encompassing everything from real-time traffic conditions and road quality to individual driving behaviors and city-wide events. Web3's decentralized framework is poised to redefine how this data is managed and utilized, shifting the focus towards enhanced service delivery and novel ownership incentives (Liu et al., 2024). This transformation promises to optimize urban mobility, personalize user experiences, and revolutionize predictive maintenance, all while empowering users with greater control over their own data.

AutoQDrive exemplifies how Web3 technologies transform the smart mobility landscape by decentralizing governance, enhancing user control, and promoting

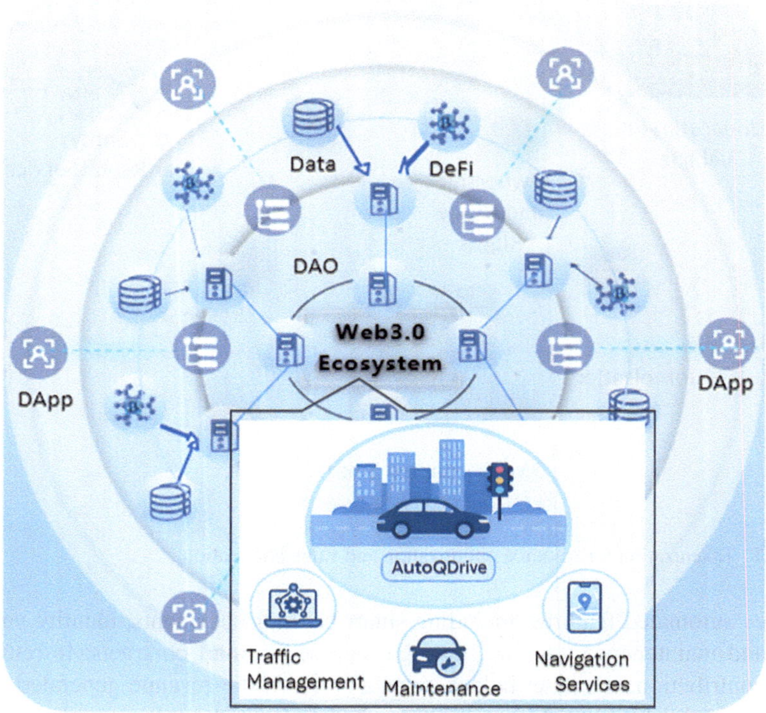

Fig. 4.2 Impact illustration through three key service improvements

data-driven optimization. Three key service improvements (Fig. 4.2) can be illustrated below:

- Intelligent Traffic Management: *AutoQDrive* aggregates and anonymizes real-time data from connected vehicles to optimize urban traffic flow. A DAO, composed of city planners, traffic engineers, citizens, and local stakeholders (e.g., road service providers), collectively oversees the system. Using smart contracts, the DAO dynamically adjusts traffic signal timing based on congestion levels, significantly reducing delays and emissions. Additionally, data on road quality, such as real-time reports of potholes or construction zones, help corresponding city planners and road service providers prioritize maintenance and allocate resources efficiently.
- Personalized Navigation Services: Within *AutoQDrive*, drivers access a DApp that offers real-time, context-aware routing. It integrates live traffic data, weather forecasts, parking availability, and public transit schedules to provide real-time recommendations for optimal routes. Leveraging DIDs, drivers maintain full control over their data, choosing what to share and with whom, ensuring privacy while benefiting from personalized experiences.

- Predictive Maintenance Optimization: With user consent, vehicle sensor data is securely shared to anticipate potential vehicle issues before they become critical. Automakers and service providers can then offer proactive maintenance suggestions, helping reduce the likelihood of breakdowns, improve safety, and extend vehicle lifespans.

In QCity's Web3-empowered mobility platform, *AutoQDrive*, data is not just a resource; it is a shared asset co-created and governed by the community. Through mechanisms such as tokenized rewards, DID systems, and DAO governance, stakeholders are empowered to contribute, control, and co-manage data in a secure, equitable, and incentivized manner:

- Tokenized Rewards: To encourage data sharing, *AutoQDrive* rewards drivers and service providers with QCoins, a native cryptocurrency in DeFi, for contributing anonymized vehicle and mobility data. These tokens can be redeemed for mobility-related services across QCity, including toll payments, EV charging credits, smart parking access, and public transit fares, creating a closed-loop incentive model that reinforces participation and value co-creation.
- Data Sovereignty with DIDs: Leveraging DIDs and smart contracts, all stakeholders, such as drivers, infrastructure providers, service developers, and public entities, maintain granular control over their own data. Each participant can define what specific data is shared, under what conditions, with whom, and for what purposes, ensuring privacy, transparency, and ethical data use.
- DAO Governance: The *AutoQDrive* DAO serves as a participatory governance framework, enabling stakeholders to propose, vote on, and influence the platform's future direction. Decisions may include the development of new infrastructure (e.g., bike lanes, charging stations, and hubs), modifications to data-sharing protocols, or reinvesting token reserves. This model transforms users from passive data subjects into active co-governors of the ecosystem. A specific example will be discussed in the following subsection on how DAO-enabled participation further enhances decision-making across QCity's mobility landscape.

AutoQDrive enhances mobility in QCity by reducing congestion, offering personalized and enhanced services, and promoting sustainable transportation. At the same time, it empowers participants by giving them control over their own data and rewarding them for their active participation. Promisingly, *AutoQDrive* facilitates the realization of a more equitable, efficient, and sustainable mobility ecosystem in QCity.

4.2.3 Stakeholders' Participation in Decision-Making

AutoQDrive DAOs offer a robust framework for integrating user participation into the governance and development of the smart car ecosystem. For example, by leveraging DIDs, participants can engage as verified stakeholders in the design and development phases of QCity's infrastructure projects, thereby fostering a more democratic, transparent, and inclusive model of urban planning and resource allocation.

More specifically, consider a proposal to develop a new smart parking facility along a busy highway to alleviate congestion and support increased EV adoption in QCity. A dedicated DAO could be established to manage this initiative, allowing stakeholders, ranging from everyday users to institutional partners, to acquire governance tokens and participate in key decision-making processes. These may include voting on design elements, determining the types of infrastructure to include (e.g., EV charging stations, IoT-enabled parking sensors), and deciding how project funds are allocated (Fig. 4.3).

Such a collaborative governance model ensures that the value of the development is co-created by all involved participants, including local governments, urban developers, and the citizens who will ultimately benefit from the infrastructure. It transitions infrastructure development from a traditional top-down approach to a community-driven model, enhancing responsiveness to citizens' needs, improving transparency, and increasing societal impact. By integrating DAOs into the smart mobility landscape, the ecosystem can enable and foster a more participatory,

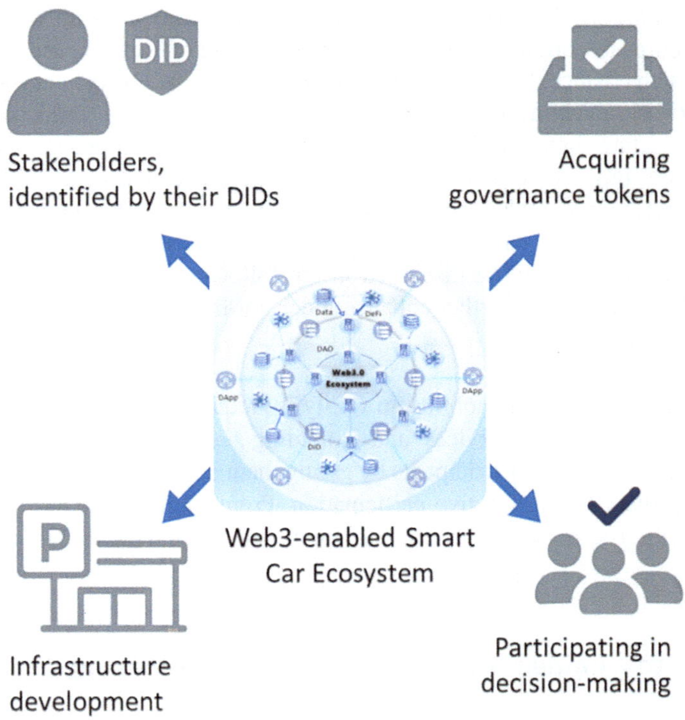

Fig. 4.3 Stakeholders' participation in decision-making and value co-creation

equitable, and resilient form of governance, fully aligned with the Web3 ethos of decentralized innovation.

4.3 Navigating Challenges in Web3-Driven Value Co-creation for the Smart Car Ecosystem

Web3, the decentralized web, represents the next stage in the evolution of the Internet. As elaborated earlier, it leverages blockchain and privacy-preserving computation technologies to power decentralized applications and enable peer-to-peer interactions, thereby departing from traditional platforms by promoting innovative methods and tools to empower participants to manage their own digital identities and information assets. This shift from Web2's platform-centric model focuses on delivering a more secure, transparent, and user-centric digital experience.

As a case study above, an intelligent, participatory, and equitable mobility ecosystem, such as the *AutoQDrive* platform, can be fully designed, developed, deployed, and realistically operational when Web3 is well-integrated. By aligning with Web3's core principles of user empowerment and ethical data stewardship, the deployed ecosystem can convert data into a shared, co-created, and stakeholder-governed asset. Specifically, the scenarios from *AutoQDrive* show how data from connected vehicles, public infrastructure, and smart mobility systems can be transformed into actionable intelligence while preserving user control and consent. Attributed to the embedded foundational functions in DIDs, DApps, and DAOs, such an ecosystem truly meets the promises by fostering value co-creation, efficient traffic flow, proactive maintenance, and incentivized data sharing through tokenized rewards.

However, to unlock the full potential of Web3 in support of the smart car ecosystem over time, as illustrated by platforms like *AutoQDrive*, the proposed framework must support quick adaptation and scaling that is both responsible and inclusive Several technical challenges must be addressed to ensure resilience, fairness, adaptability, and scalability, thereby retaining future growth and applicability in the face of changing technology, business, and societal needs. The remaining discussion in this section outlines five key challenge areas along with strategic, forward-looking solutions designed to support adaptable and scalable Web3-driven value co-creation within the smart car ecosystem in the foreseeable future.

4.3.1 Data Privacy and User Trust

The aggregation of sensitive data, such as owner identities, precise geolocation, driving patterns, and vehicle diagnostics, raises considerable concerns around privacy and surveillance. Without implementing rigorous personal data safeguards, user trust and confidence can be eroded, undermining the ecosystem's foundation and long-term sustainability (Krause, 2024).

Building user trust and confidence, while ensuring data privacy, is paramount in the ever-changing digital world. The following solutions provide a robust framework for addressing these critical issues:

- Hybrid Architectures and Privacy-Preserving Computation: Implement hybrid ecosystem architectures that strategically combine on-chain governance mechanisms with off-chain privacy-preserving computation techniques. This includes the adoption of federated learning, which enables collaborative model training without direct data sharing, and zero-knowledge proofs, allowing for data verification without revealing the underlying personal digital identities and information.
- Granular Consent Management: Empower users with fine-grained control over their own data using DIDs and audited smart contracts. Unlike traditional systems, where users have little control over how their data is accessed, used, and shared, this approach allows individuals to precisely define how their data is collected, processed, and utilized. This not only ensures transparency but also upholds user autonomy, giving users true ownership of their digital identities and information assets.

4.3.2 Sustainable Tokenomics and Ecosystem Stability

The design and implementation of tokenomic systems within Web3-empowered smart car ecosystems necessitate careful planning to prevent detrimental outcomes, such as inflation, speculative bubbles, and ecosystem fragmentation (Tan, 2023).

To address these issues related to sustainable tokenomics and ecosystem stability, the following solutions can be applied:

- Deflationary Mechanisms: Integrate various deflationary mechanisms into the token economy, such as token burning upon service redemption, to effectively manage token supply and promote long-term value stability.
- Utility-Driven Token Value Scheme: Practically establish a strong correlation between token utility and tangible ecosystem growth metrics. Examples include tying token value to increased service utilization within the platform or a measurable reduction in traffic congestion within the supported urban areas.
- Tiered Rewarding or Incentive Structures: Develop tiered rewarding or incentive programs that prioritize and reward long-term contributions and active participation over short-term speculative activities, fostering a healthy, well-engaged, and sustainable ecosystem.

4.3.3 Interoperability Across Systems and Standards

The potential for fragmented platforms and isolated, proprietary data silos poses a significant challenge to achieving seamless and comprehensive ecosystem-wide integration (Park et al., 2023; SAFIT, 2025).

To address these issues related to interoperability across systems and standards, the following solutions can be applied:

- Cross-Chain Protocols: Adopt cross-chain protocols and technologies (e.g., Loom Network, Cosmos, Plasma, Lisk, and ChainWeaver) to enable interoperability between diverse mobility platforms, IoT networks, public data repositories, and blockchains.
- Open Data Standards and Application Programming Interfaces (APIs): Actively promote the development and adoption of open data standards and APIs to support seamless data integration and sharing across different systems while rigorously preserving user data sovereignty, privacy, and access control.

4.3.4 Inclusive and Efficient DAO Governance

Participation in DAOs can be hindered by factors such as voter apathy, where members lack motivation to participate, or the potential for governance manipulation by dominant stakeholders who hold a disproportionate amount of voting power (Li & Chen, 2024).

To address these issues related to inclusive and efficient DAO governance, the following solutions can be applied:

- Weighted Voting Systems: Design and implement weighted voting systems that incorporate multiple factors beyond just token ownership, such as reputation scores reflecting contributions, loyalty, and domain expertise. This will reward genuine engagement and expertise rather than just financial power to ensure a more balanced and respected representation of stakeholder interests within the ecosystem.
- Advanced Governance Models: Incorporate more advanced governance models, such as delegated voting, where users can delegate their voting power to trusted representatives, and quadratic voting, which aims to give more weight to individual preferences and minority voices when appropriate, thereby fostering promised inclusivity.
- Community Engagement Incentives: Implement mechanisms to actively encourage community engagement in governance processes. This can include rewarding participation in proposal development and discussions, as well as allocating tokens or voting rights, as mentioned above, for active involvement in decision-making processes.

4.3.5 Technical Complexity and User Accessibility

The inherent technical complexity associated with emerging Web3 technologies, particularly in the development and usage of the Web3 ecosystem, can present a significant barrier to widespread adoption when intuitive or user-friendly interfaces are lacking. This can discourage potential developers or users from getting involved, as they may find the learning curve too steep (Bobrova & Perego, 2025).

To address these issues related to technical complexity and user accessibility, the following solutions can be applied:

- User-Friendly Interfaces and DApps: Prioritize the development of intuitive user interfaces and mobile-first DApps with streamlined onboarding processes to minimize the potential learning burden and enhance user experience.
- Educational Resources and Support: Provide comprehensive educational resources and readily accessible support channels to demystify blockchain technology, smart contracts, token economics, and other core Web3 concepts for users with varying levels of technical backgrounds.
- Human-Centered Design: Integrate human-centered design principles throughout the development process, with a strong emphasis on usability, accessibility, and inclusivity, to ensure the platform caters to the diverse needs and preferences of its user base.

4.4 Toward a Resilient, Equitable, and Sustainable Mobility Future

When the principles of Service Science are well adopted, the technical integration of Web3 into the smart car ecosystem will surely unlock a collaborative value loop, where all participants, including users, infrastructure operators, developers, investors, and policymakers, co-design and co-create mobility services, collectively optimize city mobility systems, and benefit equitably from the enabled transparent, user-driven, and digital mobility networks. As revealed in this case study, by decentralizing data governance and aligning incentives with broader user needs or societal goals, Web3-empowered platforms like *AutoQDrive* are not only enhancing city mobility services; they are redefining the very foundations of trust, participation, governance, and innovation in the future digital cities. Evidently, this shift aligns perfectly with Web3's core principles of user empowerment through DIDs, decentralized governance via DAOs, ethical data stewardship, and monetization that benefits all participants. The fundamental change clearly elaborates that data is treated as a shared asset rather than a corporate commodity.

As comprehensively explored in this chapter, only by addressing the multifaceted challenges outlined above through strategic, inclusive, and user-centric solutions can Web3-empowered platforms realize their full transformative potential. This involves addressing issues such as data privacy with hybrid architectures and granular consent management, ensuring sustainable tokenomics to prevent ecosystem instability, promoting interoperability across fragmented systems, fostering inclusive and efficient DAO governance, and overcoming technical complexity to ensure user ownership and enhance user accessibility. In conclusion, by leveraging DAO, DID, DeFi, DApp, and other emerging Web3 functional services, the convergence of decentralized governance, tokenized incentives, and data sovereignty reimagines smart mobility not just as a technological upgrade but as a resilient, ethical, and participatory city mobility services designed for the future of urban life, unprecedentedly contributing to smarter, safer, and more sustainable human living environments.

References

Abirami, S., Pethuraj, M., Uthayakumar, M., & Chitra, P. (2024). A systematic survey on big data and artificial intelligence algorithms for intelligent transportation system. *Case Studies On Transport Policy, 17*, 101247.

Alam, T. (2024). Data privacy and security in autonomous connected vehicles in smart city environment. *Big Data and Cognitive Computing, 8*(9), 95.

Alam, T., Gupta, R., Nasurudeen Ahamed, N., Ullah, A., & Almaghthwi, A. (2024). Smart mobility adoption in sustainable smart cities to establish a growing ecosystem: Challenges and opportunities. *MRS Energy & Sustainability, 11*(2), 304–316.

Alexakis, T., Peppes, N., Demestichas, K., & Adamopoulou, E. (2022). A distributed big data analytics architecture for vehicle sensor data. *Sensors, 23*(1), 357.

Bittencourt, J. C. N., Costa, D. G., Portugal, P., & Vasques, F. (2024). A survey on adaptive smart Urban systems. *IEEE Access, 12*, 102826–102850.

Bobrova, P., & Perego, P. (2025). The development of user-centric design guidelines for Web3 applications: An empirical study. *Computers, 14*(2), 46.

Butler, L., Yigitcanlar, T., & Paz, A. (2020). Smart urban mobility innovations: A comprehensive review and evaluation. *IEEE Access, 8*, 196034–196049.

Damaj, I. W., Yousafzai, J. K., & Mouftah, H. T. (2022). Future trends in connected and autonomous vehicles: Enabling communications and processing technologies. *IEEE Access, 10*, 42334–42345.

Dong, M., & Qiu, R. G. (2004). An approach to the design and development of an intelligent highway point-of-need information service system. In *Proceedings. The 7th International IEEE Conference on Intelligent Transportation Systems* (pp. 673–678). IEEE.

Ghosh, D., Agrawal, A., & Sundararajan, M. (2022). OTA updates and the future of automotive software. *Journal of Software Engineering for Automotive Systems.*

Khan, M. A., Ghosh, S., Busari, S. A., Huq, K. M. S., Dagiuklas, T., Mumtaz, S., et al. (2021). Robust, resilient and reliable architecture for v2x communications. *IEEE Transactions on Intelligent Transportation Systems, 22*(7), 4414–4430.

Krause, D. (2024). Web3 and the decentralized future: Exploring data ownership, privacy, and Blockchain infrastructure. *Privacy, and Blockchain Infrastructure (December 19, 2024).*

Li, S., & Chen, Y. (2024). Governing decentralized autonomous organizations as digital commons. *Journal of Business Venturing Insights, 21*, e00450.

Liu, W., Cao, B., & Peng, M. (2024). Web3 technologies: Challenges and opportunities. *IEEE Network, 38*(3), 187–193.

Lonergan, K. (2021). *The Evolution of Automobiles*. Retrieved on April 5, 2025, from https://opentextbooks.clemson.edu/sts1010fidlerfall2021/chapter/the-evolution-of-automobiles/

Martinho, A., Herber, N., Kroesen, M., & Chorus, C. (2021). Ethical issues in focus by the autonomous vehicles industry. *Transport Reviews, 41*(5), 556–577.

Park, A., Wilson, M., Robson, K., Demetis, D., & Kietzmann, J. (2023). Interoperability: Our exciting and terrifying Web3 future. *Business Horizons, 66*(4), 529–541.

Qiu, R. G. (2024). Web3 and service science: Perfectly aligned to maximize the values of all stakeholders in the digital economy. *Digital Transformation and Society, 3*(4), 337–339.

Qiu, R. G. (2014). *Service science: The Foundations of service engineering and management.* Wiley.

Razi, Q., Piyush, R., Chakrabarti, A., Singh, A., Hassija, V., & Chalapathi, G. S. S. (2025). Enhancing data privacy: A comprehensive survey of privacy- enabling technologies. *IEEE Access*.

Romeu Lezama, J. J. (2016). *Architectural innovation in the automotive industry: Tesla and the renaissance of the battery electric vehicle* (Master Thesis). Massachusetts Institute of Technology.

SAFIT. (2025). *ChainWeaver: Weaving chains into a network and sharing connectivity*. Retrieved January 10, 2025, https://docs.chainweaver.org.cn/index.html

Shrivastava, A., Jangalwa, M., & Vyas, S. (2024). A decentralized application for car rentals: Towards Blockchain application. In *2024 IEEE 4th international conference on ICT in business industry & government (ICTBIG)* (pp. 1–7). IEEE.

Shanmugam, K., Rana, M. E., & Hong, F. T. Y. (2024). Autonomous intelligent vehicles: Impact, current market, future trends, challenges, and limitations. *Optimized Computational Intelligence Driven Decision-Making: Theory, Application and Challenges*, 173–194.

Singh, B., Vig, K., Dutta, P. K., & Kaunert, C. (2025). Illuminate metaverse multisensor fusion and dynamic routing technologies across Web3-powered for autonomous vehicles shaping efficient Urban transport solutions of future in smart city era: Deep dive into protocols for benefiting society lensing prospects and challenges. *Building Embodied AI Systems: The Agents, the Architecture Principles, Challenges, and Application Domains*, 395–416.

Souri, A., Zarei, M., Hemmati, A., & Gao, M. (2024). A systematic literature review of vehicular connectivity and V2X communications: Technical aspects and new challenges. *International Journal of Communication Systems, 37*(10), e5780.

Tan, L. J. (2023). Token economics model creation and security. In *A comprehensive guide for Web3 security: From technology, economic and legal aspects* (pp. 103–117). Cham.

Van Den Boom, F. (2022). *Driven by digital innovations: Regulating connected car data access and use, for Telematics Insurance in Europe* (Doctoral dissertation). Bournemouth University.

Zhao, D., Wang, Q., & Wang, H. (2020). *AI technologies for autonomous driving*. ACM Computing Surveys.

5 Value Co-creation Case Study: The Smart City Ecosystem

Abstract

In this chapter, we present a case study of the emerging smart city ecosystem. With rapid global urbanization projected to place nearly 70% of the world's population in cities by 2050, urban centers face mounting challenges in infrastructure, resource management, safety, and social equity. We explore how Web3 technologies, including Decentralized Autonomous Organizations (DAOs), Decentralized Identifiers (DIDs), decentralized finance (DeFi), and decentralized applications (DApps), can address these issues by replacing traditional top-down governance with a decentralized, value co-creation framework. More importantly, using the conceptual *QCity Eco* platform, built on the pillars of Web3, we demonstrate three transformative value co-creation scenarios. First, asset digitization tokenizes individual-owned resources (e.g., solar panels) into NFTs managed through DeFi, advancing sustainability and generating creative revenue streams. Second, community-driven governance and tokenized incentives encourage citizens to share their diverse data sources, transforming them into collective-value assets that enable intelligent city services, foster participatory governance, and support data-driven urban planning. Third, collaborative decision-making fosters inclusivity by empowering all stakeholders, including individual citizens, to co-design and collectively shape the city's future.

Keywords

Smart city ecosystem · Tokenized rewards · Asset digitization · Data value · Resource utilization · Participatory decision-making

R. Qiu, *Value Co-Creation in Web 3.0*, Advances in Data Analytics, AI, and Smart Systems, https://doi.org/10.1007/978-3-032-12911-6_5

5.1 Introduction to the Smart City Ecosystem

Urbanization has been a defining force in shaping human civilization for over 5000 years. The first recorded urban area in human history was Mesopotamia, according to Reba et al. (2016). At around 3000 BCE, more cities gradually arose in other fertile river valleys, such as the Indus Valley, the Yellow River, and Ancient Egypt, and later quickly flourished due to agricultural surpluses and burgeoning trade. These early urban centers became hubs of administration, commerce, culture, and innovation for millennia.

The Industrial Revolution of the eighteenth and nineteenth centuries brought about a profound shift in human civilization. Driven by innovations in mechanized production, improved transportation, and a fundamental economic shift away from agriculture, as well as the strengthening of the manufacturing industry, this era catalyzed a massive migration of rural populations into booming industrial cities worldwide. As a result, urban growth exploded, with the global population living in cities reaching approximately 15% by 1900 (Reba et al., 2016).

Today, less than 45% of the global population lives in rural areas, and the trend of urbanization continues to accelerate. United Nations (UN) projections suggest that by 2050, this figure will shrink to nearly 30% (UN, 2018). In other words, about 70% of the global population will live in urban areas. In fact, this unprecedented demographic shift is already creating a new reality defined by megacities such as Tokyo, Delhi, São Paulo, Shanghai, and Beijing, each of which houses tens of millions of residents.

This accelerated urban expansion has put immense pressure on existing city infrastructure, governance systems, and natural resources (Humbal et al., 2023; UN, 2018). Currently, continuous urbanization grapples with a series of persistent and interconnected challenges, as highlighted below:

- Overcrowding and Urban Sprawl: Unregulated expansion leads to slums, inadequate housing, and traffic congestion.
- Pollution: Cities face alarming levels of air, water, and land contamination (e.g., Delhi's toxic smog, Lagos' overflowing landfills).
- Resource Strain: Unsustainable energy consumption, water shortages, and inefficient waste management systems.
- Climate Change: Urban areas account for 70% of CO_2 emissions and are highly vulnerable to climate-related risks, including heatwaves, floods, and rising sea levels.
- Social Inequity: Deepening digital divides, lack of access to basic services, and increased urban poverty.
- Governance Complexity: Traditional bureaucratic systems struggle to respond dynamically to fast-evolving urban demands.

To tackle these pressing issues, the concept of the smart city has emerged. The development of smart cities is not just a single advanced technology or updated city blueprint, but rather a dynamic, evolving solution to urban innovation, resilience, and sustainability. It defines a new urban paradigm that has become a defining force in contemporary city development, essentially creating a better living environment for human beings (Anthopoulos, 2017).

The smart city framework envisions a future of enhanced city operational efficiency and improved management of urban activities and utilities, made possible by significant technological advancements and innovation over the last several decades (Komninos et al., 2022). According to Anthopoulos (2017), early initiatives in the 1990s were largely experimental, focusing on simple data visualization, such as digital city maps or community portals. As digital infrastructure evolved, however, so did the concept itself, transitioning from basic information platforms to interactive systems for real-time urban service delivery and citizen engagement.

According to Anthopoulos (2017), the smart city evolution can be divided into key phases:

- Digital Cities (1990s–2000s): Emphasized web-based information sharing and virtual simulations of physical cities for improved city management (e.g., Digital Kyoto) (Ishida, 2002).
- Ubiquitous Cities (2000s): Introduced sensor networks and real-time data to embed intelligence into physical infrastructure for improved city safety, control, and management (Anthopoulos & Fitsilis, 2010).
- Smart Cities (2010s–present): Focus on cyber-physical systems, sustainability, open innovation, and citizen-centric governance for smart living (Silva et al., 2018).

This evolution was also paralleled by the growing recognition of smart cities as socio-technical ecosystems, where infrastructure, policy, people, data, and technology intersect, well aligned with the principles of Service Science (Qiu, 2014).

However, modern urbanization has resulted in complex pressures on human beings, including rapid population growth, environmental degradation, and infrastructural strain. It is the smart city that has become a strategic response to those complex challenges. Currently, the core smart city relies on technologies that create a living urban nervous system. For example, the Internet of Things (IoT) enables objects to be sensed and managed through the Internet, promoting and supporting a coherent integration between the physical world, software applications, and human beings. This physical sensing is complemented by social sensing, where mobile apps provide real-time insights into our living communities and environments (Qiu et al., 2019).

Such an intricate network requires robust governance and positive citizen participation for better reliability and healthy growth. Indeed, it is crucial to involve citizens in the development process to promote sustainability (Caputo et al., 2023). The success of a smart city is heavily influenced by the strategic decisions and interactions of diverse stakeholders (Kociuba et al., 2023). However, this interconnected vision faces a major obstacle: the compartmentalization of solutions across different sectors. This lack of interoperability and resource sharing undermines the very goals of smart cities, limiting their overall impact on efficiency, cost reduction, and

CO_2 emissions (Komninos et al., 2022). Unlocking the ecosystem's full potential requires moving beyond these centered, siloed approaches and toward a truly integrated and collaborative urban framework.

In response to the complex challenges of twenty-first century urbanization, the smart city ecosystem has emerged as a strategic solution. It must integrate information technology, networking, big data, IoT, AI, and cloud computing with sustainable urban planning and participatory governance (Pandey et al., 2018). For example, integrating diverse electronic data sources into a unified platform is crucial for smart city development, requiring a robust and scalable framework for integrating big data and analytics solutions to manage city services effectively. Digital twin frameworks have also emerged to enhance smart city functions, particularly in areas such as smart mobility (Faliagka et al., 2024). Developing "Connected Intelligence Spaces" has great potential for creating high-impact smart cities. These spaces facilitate cross-sector collaboration, effective data utilization, and value co-creation among all stakeholders, thereby evolving the smart city ecosystem (Komninos et al., 2022; Qiu, 2024).

Key drivers of this ecosystem evolution include the following:

- Technological innovation: Real-time monitoring (e.g., IoT), predictive analytics (e.g., AI), and seamless communication (e.g., 5G/6G) have transformed how cities manage energy, traffic, safety, and health (Qiu et al., 2019) and the emerging Web3 as elaborated in the last two chapters.
- Policy and sustainability mandates: Driven by international frameworks such as the UN Sustainable Development Goals (SDGs) (Estevez et al., 2021) and the Paris Agreement, cities are now strategically prioritizing low-carbon, resilient development.
- Collaborative governance: Public-private partnerships and citizen engagement platforms have reshaped urban planning and service delivery (Paskaleva et al., 2021).
- Equity and resilience: Smart ecosystems aim to bridge gaps in access and opportunity (James et al., 2021).

The smart city ecosystem is not a mere technological enhancement, but a fundamental reimagining of urban life. It requires a holistic, systems-based approach that seamlessly integrates critical urban systems, such as transportation, energy, water supply, waste management, healthcare, and city governance, utilizing technologies like IoT, big data analytics, and AI and machine learning (Pandey et al., 2018). At its core, this evolving paradigm places citizens at the center, rather than technologies, in a classic smart city, utilizing digital tools as enablers of a more responsive, inclusive, and sustainable urban environment (Caputo et al., 2023). Today and in the future, its real power lies in balancing innovation with human-centered governance and design, ensuring that as cities evolve, they do so in ways that are equitable, adaptive, and deeply livable.

Notably, the smart city ecosystem is multifaceted, requiring a balanced approach across several key dimensions to achieve its full potential. Fundamentals among these are the perspectives of mobility, safety, and sustainability with strong and effective governance. For example, smart city mobility initiatives seek to redefine urban mobility through innovative transport solutions and traffic management (Pandey et al., 2018; Qiu et al., 2019), while ensuring the safety and security of citizens through advanced surveillance and emergency response systems (Anthopoulos, 2017; Komninos et al., 2022). Simultaneously, sustainability imperatives drive the integration of green technologies and resource-efficient practices (Caputo et al., 2023; Kociuba et al., 2023). Effective governance frameworks are crucial for promoting citizen engagement, stakeholder collaboration, and the delivery of equitable services (Kociuba et al., 2023). The following sections will delve into each of these three perspectives, exploring their significance within the smart city ecosystem and highlighting examples of their implementations.

5.1.1 The Mobility Perspective

Smart cities utilize technologies to transform the movement of people and goods within the urban environment. This transformation encompasses optimizing traffic flow, promoting efficient public transport, and exploring innovative solutions for shared mobility, all of which are crucial for the development of smart cities (Anthopoulos, 2017). As well explored in Chap. 4, the integration of physical and social sensing, for example, enhances smart city mobility services, providing a more comprehensive understanding of urban mobility dynamics (Qiu et al., 2019).

This enhanced understanding enables the development of a new generation of data-driven mobility services. Cities can implement dynamic traffic management systems that adjust signal timings in real time to alleviate congestion. Similarly, smart routing algorithms provide commuters with personalized, multimodal travel options that seamlessly combine public transit, shared bikes, and ride-hailing services. Ultimately, these technologies aim to create a more seamless and personalized travel experience, reducing commute times and minimizing the environmental footprint of transportation. Looking ahead, the integration of autonomous vehicles and intelligent logistics networks promises to further transform urban mobility, moving toward a future of Mobility as a Service (MaaS), where citizens can access a full suite of transport options on demand.

5.1.1.1 Traffic, Road Safety, and Parking Management

In the intelligent traffic management center, a multitude of network-based sensors are deployed to monitor detailed traffic flow information. As largely explored in Chap. 4, these sensors, installed on road surfaces, lamp posts, traffic signs, vehicles, or cellular phones, directly or indirectly capture real-time data on vehicle types, speeds, volumes, and congestion levels. By analyzing these data, traffic management departments can dynamically adjust traffic light timings to optimize traffic flow. For example, during rush hour, green lights can stay on longer on busy roads, allowing more cars to get through and reducing congestion.

Vehicle-to-Vehicle (V2V) and Vehicle-to-Infrastructure (V2I) communication technologies are pivotal to modern urban mobility, catalyzing great improvements in traffic control efficiency and public safety. Intelligent connectivity, a key component in modern smart vehicles, is paramount, enabling real-time data exchange across the mobility landscape. V2V communication enables vehicles to exchange information, such as speed and direction, providing early warnings of potential collision risks. V2I communication allows vehicles to receive information from traffic lights, roadside sensors, and other infrastructure, aiding in route planning by providing updates on road construction or congestion (Anthopoulos, 2017; Souri et al., 2024).

Intelligent parking management systems are integral to traffic management (Anthopoulos, 2017). Sensors installed in parking lots and roadside spaces provide real-time occupancy data, which is relayed to drivers in real time. Mobile applications streamline the urban parking experience by providing real-time availability data, thereby directly reducing time searching for parking spaces and alleviating traffic congestion.

The concept of smart mobility, its ecosystem as discussed in Chap. 4, extends to connected vehicles, which are a key component of the smart car ecosystem. These vehicles rely on a comprehensive suite of IoT sensors, GPS, and robust cellular/Wi-Fi connectivity to enable real-time data exchange. This interconnectedness, encompassing V2V, V2I, and Vehicle-to-Everything (V2X) communication, is crucial for enhanced safety, optimized traffic flow, and the delivery of real-time information. Enhancing road efficiency, control, and safety highlights the growing importance of technology-driven solutions in urban mobility services (Pulkkinen et al., 2019; Touloumidis et al., 2024).

5.1.1.2 Public Transportation

Intelligent bus systems are becoming a foundational part of modern public transportation (Rong et al., 2022). Equipped with GPS tracking and advanced dispatching algorithms, these systems enable real-time monitoring of vehicle locations and operational conditions, allowing for dynamic route adjustments and optimized scheduling. For instance, during peak demand periods, additional buses can be deployed on high-traffic routes to minimize congestion and reduce passenger wait times. Complementing this, smart bus stops now provide real-time arrival updates and are increasingly outfitted with user-friendly amenities such as electronic fare payment systems, Wi-Fi connectivity, and heated seating, significantly enhancing the overall commuter mobility experience.

MaaS platforms elevate urban mobility by consolidating multiple transport modes, including buses, trains, and ride-hailing, into a unified, app-based ecosystem. These platforms facilitate seamless trip planning, booking, and payment, fundamentally changing how citizens navigate their cities (Jittrapirom et al., 2017). Simultaneously, ride-sharing and carpool services, supported by the sharing economy, offer on-demand, flexible mobility options. These services are not only convenient and accessible via the app but also intelligently managed: AI-driven systems optimize the distribution of shared vehicles according to real-time user demand and urban mobility patterns.

Most critically, the application of big data analytics and artificial intelligence has revolutionized public transportation operations, enabling predictive maintenance, real-time demand forecasting, and adaptive scheduling, all of which contribute to heightened efficiency and service reliability (Alam et al., 2024; Touloumidis et al., 2024).

Public transportation thus serves as the backbone of smart urban mobility, enabling cities to be more equitable, sustainable, and resilient. Its ongoing transformation into an intelligent, interconnected, and human-centered system reflects the broader imperative to integrate technology to build smart city ecosystems that are livable, inclusive, sustainable, and future-ready.

5.1.2 The Safety Perspective

Smart city ecosystems prioritize the safety and resilience of urban environments through integrated systems that encompass both proactive surveillance and security measures, as well as reactive emergency response capabilities (Anthopoulos, 2017; Troisi et al., 2022):

- Surveillance and Security: Smart cities are moving beyond traditional surveillance, which focuses solely on criminal activity. They are adopting comprehensive systems that utilize high-definition camera networks, AI-driven video analysis, and IoT sensor networks to monitor a range of urban dynamics, including traffic flow, public order, and potential security threats. These technologies enable real-time detection of anomalies, suspicious activities, and environmental hazards, supporting both crime prevention and overall urban management. Protecting the data generated by these systems is critical, necessitating robust encryption and security protocols.
- Emergency Response: Smart cities are also developing advanced emergency response mechanisms to handle events such as fires, natural disasters, and other crises. These systems rely on smart sensors for early detection, intelligent traffic management for optimized emergency vehicle routing, and communication platforms to disseminate information and coordinate response efforts. The goal is to minimize damage, protect citizens, and ensure efficient resource allocation during emergencies.

This integrated approach to safety, combining proactive surveillance with responsive emergency systems, is a defining characteristic of future smart city development.

5.1.2.1 Surveillance and Security

Smart cities are increasingly adopting comprehensive surveillance and security measures to ensure the safety and resilience of urban environments. This involves a notable shift from traditional surveillance models, which primarily focused on criminal activity, to a broader, more integrated approach that incorporates a diverse set of urban monitoring functions.

In a smart city ecosystem, surveillance extends to various critical areas, including overseeing traffic violations, managing environmental pollution incidents, ensuring public safety, and maintaining order during large gatherings or emergency events. This expanded scope underscores a commitment to real-time governance and responses, enhancing a city's ability to effectively address a wide range of unexpected urban challenges.

Several key technologies and systems drive this enhanced surveillance and security framework (Anthopoulos, 2017; Troisi et al., 2022). High-definition camera networks, equipped with sophisticated image recognition and machine learning technologies, are strategically deployed across urban areas to detect deviations from normal patterns. For instance, these systems can distinguish between standard traffic flow and anomalies, such as congestion, accidents, or unauthorized public gatherings. The integration of AI plays a crucial role in enabling the automated interpretation of video feeds (Zahra et al., 2024), which facilitates the timely issuance of alerts and the rapid deployment of response teams. Furthermore, intelligent video analysis systems are employed to enhance security by identifying suspicious behavior within public spaces, such as prolonged loitering, trespassing in restricted areas, or the abandonment of objects in sensitive zones, thus supporting a preventive security posture through early intervention. IoTs also form a fundamental layer of this surveillance infrastructure, with a distributed network of sensors, including motion detectors, vibration sensors, and acoustic monitors, embedded throughout public and private spaces to enable real-time detection of physical security threats like unauthorized entry or structural disturbances, automatically triggering alarms and notifications to security control centers.

Given the vast volume of data generated by surveillance systems in smart cities, ensuring data security and privacy is of critical importance. Protecting sensitive information, such as individual movements, public infrastructure details, and urban operational data, requires the implementation of robust encryption, secure storage architectures, and stringent access control mechanisms. Ultimately, safeguarding the confidentiality and integrity of this data is not only a technical imperative but also essential for maintaining public trust and supporting the ethical and sustainable operation of smart city ecosystems.

5.1.2.2 Emergency Response

Smart city ecosystems are increasingly designed to support rapid, coordinated, and adaptive emergency response capabilities, enabling cities to address a broad spectrum of unpredictable events, including fires, natural disasters, industrial accidents, and acts of terrorism. By leveraging advanced technologies such as IoTs, AI, and big data analytics, these systems enhance real-time detection, situational assessment, and crisis management (Costa et al., 2022; Abraham et al., 2024; Alam et al., 2024). By integrating these technologies into the urban fabric, cities aim not only to minimize the impact of emergencies but also to enhance overall resilience and the well-being of residents through more responsive and sustainable public services.

Smart sensors play a foundational role in the early detection of hazardous events. For example, in the event of a fire, smoke and heat detectors can rapidly identify anomalies and transmit real-time alerts to the city's central emergency management system. This early warning capability enables swift intervention, reducing the potential for escalation and mitigating damage. In parallel, the smart traffic management infrastructure is equipped to dynamically adjust traffic patterns to prioritize emergency response. Traffic signal systems can be programmed to create "green corridors," ensuring that emergency vehicles are granted expedited, unobstructed access to the affected area (Qiu et al., 2019; Faliagka et al., 2024).

Moreover, intelligent communication platforms, including emergency broadcasting systems and mobile applications, disseminate real-time alerts and guidance to the public. These messages may include evacuation instructions, shelter locations, and safety protocols, empowering citizens to make informed decisions during crises (Caputo et al., 2023).

In the healthcare domain, emergency response is strengthened through integrated medical platforms that provide up-to-date information on available hospital facilities, bed occupancy, and critical care equipment. This enables authorities to allocate medical resources more efficiently during large-scale emergencies, thereby improving patient outcomes and reducing strain on healthcare systems (Abraham et al., 2024; Pandey et al., 2018).

Collectively, these innovations contribute to a more agile and resilient urban emergency response framework that prioritizes both public safety and systemic sustainability.

5.1.3 The Sustainability Perspective

Smart cities are undergoing a significant transformation to place sustainability at the heart of their operational frameworks. Central to this shift is a focus on improving energy efficiency and reducing waste to optimize resource use across metropolitan areas. This evolution is driven by interconnected technologies, including smart grids, intelligent building management systems, and data-driven waste management platforms, which work together to minimize environmental impact and promote sustainable urban development (Gupta et al., 2021; Troisi et al., 2022). The broader goal is to cultivate urban environments that not only reduce their ecological footprint but also embrace circular economy principles and promote responsible resource stewardship for future generations.

5.1.3.1 Energy Efficiency

Improving energy efficiency is a cornerstone of sustainability in smart cities. Smart grids are central to this effort, as they enable real-time monitoring, adaptive load balancing, and efficient electricity distribution. Crucially, they also promote the seamless integration of diverse renewable energy sources such as solar, wind, and hydropower, which is vital for creating a resilient, low-carbon energy infrastructure (Martins et al., 2021; Razmjoo et al., 2022; Zhou & Liu, 2024).

Intelligent building management systems further advance energy efficiency by leveraging IoT sensors and automated control technologies. These systems dynamically adjust energy consumption based on real-time occupancy data, indoor light intensity, and external environmental conditions. For instance, smart buildings utilize intelligent lighting systems that dim or intensify based on available daylight, while smart HVAC systems optimize cooling and heating by detecting temperature variations, thereby minimizing unnecessary energy consumption (Gupta et al., 2021). Moreover, intelligent monitoring of building insulation ensures optimal thermal performance across seasons.

In the transportation sector, electric vehicles (EVs) represent a crucial strategy for reducing dependence on fossil fuels. Smart cities promote EV adoption through smart charging infrastructure that adjusts charging schedules and power loads according to grid capacity. Off-peak charging, enabled by smart grids, reduces both user costs and grid strain (Jiao et al., 2020). Moreover, the incorporation of various renewable energy sources into urban power grids significantly strengthens energy sustainability. Smart grids facilitate the efficient management of decentralized energy generation and storage, ensuring a stable, adaptive, and environmentally sustainable energy supply.

5.1.3.2 Waste Reduction

Waste control is a vital constituent of sustainable city infrastructure. Intelligent systems, equipped with IoT-enabled sensors installed on waste bins, continuously monitor fill levels in real time. These systems automatically trigger alerts when bins approach capacity, enabling optimized collection routes, preventing overflows, and reducing unnecessary trips. As a result, waste collection becomes more efficient, cost-effective, and environmentally friendly, minimizing operational costs and lowering carbon emissions.

Beyond logistics, smart cities actively encourage waste classification and recycling through mobile applications and digital terminals that educate and guide citizens on proper waste segregation and disposal. Advanced image recognition and feedback systems monitor sorting behavior and provide corrective guidance to users, promoting community participation and reducing contamination rates in recyclables (Szpilko et al., 2023). Moreover, smart waste solutions, powered by big data analytics, enable predictive waste collection, dynamic allocation of collection resources, and targeted recycling initiatives. These systems significantly reduce landfill dependency and foster a circular economy, wherein resources are continually reused, refurbished, and revalorized. Smart platforms facilitate the efficient recovery of valuable materials, such as electronic components and construction debris, reinforcing responsible resource use (Bressanelli et al., 2022).

Through these innovations, smart cities are not merely optimizing existing waste management practices but are actively reshaping urban metabolism to align with long-term environmental sustainability goals.

5.2 Web3-Driven Value Co-creation in the Smart City Ecosystem

Smart cities are evolving to integrate flexibility, efficiency, and ecological responsibility into urban operations and management. This evolution is driven by digital infrastructure, green technologies, and citizen insights, resulting in adaptive, resilient, and people-centered urban ecosystems (Anthopoulos, 2017).

The success of these ecosystems depends on strong collaboration among diverse stakeholders, including governments, tech companies, infrastructure providers, service operators, and communities. This collaboration is essential for establishing governance frameworks, technical standards, and trust mechanisms (Troisi et al., 2022). Effectively addressing key areas such as mobility, safety, and sustainability requires interoperability, connectivity, and robust data security. Without this collaboration, progress can be fragmented, hindering scalability, inclusivity, and public trust.

Web3 technologies are crucial for this transformation, offering decentralization, transparency, data self-sovereignty, and interoperability. These principles enable more participatory and equitable governance and service delivery, addressing the need for citizen involvement in the core smart city domains of mobility, safety, and sustainability that were well explored above. As Qiu (2024) notes, Web3's architecture aligns with Service Science, promoting value co-creation within service ecosystems (Qiu, 2014). This alignment supports the shift from centralized, top-down city management to collaborative, citizen-empowered, and trust-driven models.

This shift extends beyond technology, driving urban innovation toward human-centered, AI-augmented, and ethically responsible development. Citizens become active co-creators, providing data, insights, and innovations to continuously improve urban life. Web3 can truly facilitate the development of smart *deepservice*, personalized, ethical, and adaptive urban services that enrich daily life (Qiu, 2024), meeting the pressing needs of the future smart city ecosystem.

In forthcoming sections, we will explore use cases that demonstrate how Web3 can facilitate value co-creation within the smart city ecosystem, advancing the development of more intelligent, resilient, and participatory cities.

5.2.1 Asset Digitization, Sharing, and Transfer to Maximize Value Co-creation

To fully realize resource sharing and maximize asset utilization within a smart city ecosystem, not only the large pool of enormous, archived data from all stakeholders or participants, but also various physical assets can be digitized, shared, and/or integrated. This process, generally known as tokenization through DeFi, involves representing tangible assets, such as vehicles, public infrastructure, or even energy units, as unique digital tokens, and intangible assets, such as collected data and enabled services. These tokens, often non-fungible tokens (NFTs), securely encode ownership and usage rights on a blockchain. For example, an underutilized public park space, a set of solar panels on a municipal building, or even an individual house can be tokenized, with their digital representation enabling traceable, verifiable, and transferable rights. These tokens can then be exchanged or utilized within a Web3-empowered smart city ecosystem, fostering a more efficient and decentralized allocation of capital and infrastructure.

Consider a city called "QCity" that has adopted a comprehensive Web3-empowered platform, named *QCity Eco*, leveraging DAO, DID, DeFi, and DApp, to manage and optimize its various smart city ecosystems. This platform integrates data from diverse sources, such as:

- IoT sensor networks across the city
- Data from connected vehicles and traffic management systems
- Energy consumption and production data
- Waste management system data
- Public health data
- Public safety and environmental monitoring systems

QCity Eco supports enhanced city services across mobility, energy, waste management, and public services, while also incentivizing citizen and stakeholder participation.

This capability aligns with Web3's emphasis on user empowerment and control over digital assets. In one scenario within QCity, the city government tokenizes a set of household solar energy generation with EV charging stations, as shown in Fig. 5.1. An energy provider acquires the NFT representing a subset of these charging stations. This entity can then leverage this digital asset to offer a decentralized EV charging service, utilizing a decentralized application (DApp) within *QCity Eco*

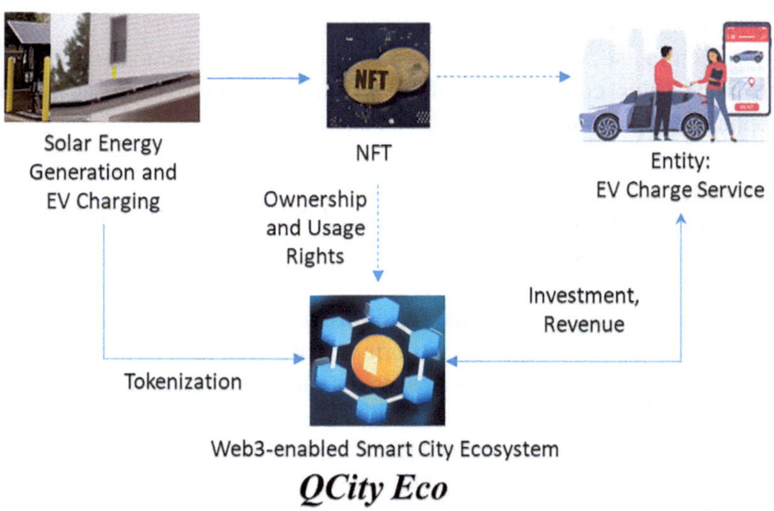

Fig. 5.1 Tokenization for resource optimization and value co-creation

for charging station management, pricing, and energy distribution. In return for their investment and service provision, the energy provider receives a share of the revenue generated from the charging stations, creating a value co-creation model.

This mechanism aligns with Web3's tenets of user empowerment, transparency, and decentralized control over digital assets. In another practical example similar to one in Fig. 5.1, individual residents in QCity could tokenize their rooftop solar panels as NFTs. Through a DApp on *QCity Eco*, they can then sell excess energy to their neighbors in a peer-to-peer energy marketplace, earning cryptocurrency tokens in return. Smart contracts automate the required energy transactions and distribution, ensuring transparency and security. The digital asset representing solar panels of a house can also be transferred to a new owner if the house is sold.

Through tokenization and Web3 platforms like *QCity Eco*, the city optimizes resource utilization, promotes citizen participation, and enables the development of innovative, decentralized services across various sectors.

5.2.2 Data-Driven Service Improvement and Ownership Incentives

Smart city ecosystems have the potential to significantly improve urban services and quality of life by leveraging data from various sources. To fully realize this potential, it is crucial to establish mechanisms that not only facilitate data sharing but also incentivize residents to actively contribute their data, fostering a sense of ownership and participation.

Enhanced Service Delivery through Data Sharing

Data sharing within a smart city ecosystem plays a pivotal role in enhancing the efficiency, responsiveness, and sustainability of urban services. A prime example is urban mobility, as discussed in Chap. 4. By integrating real-time data from connected vehicles, road sensors, and public transportation networks, cities can gain a holistic view of traffic conditions. This integrated data stream enables the identification of congestion hotspots, the optimization of traffic flow, and the strategic planning of transport route infrastructure. Informed urban planning, in turn, reduces congestion, improves air quality, and elevates the overall effectiveness of transportation systems, ultimately benefiting all citizens.

Importantly, the impact of data sharing extends far beyond mobility. Within smart cities, data generated from domains such as energy consumption, waste management, and environmental monitoring can be interconnected to improve resource efficiency and sustainability. For example, using real-time energy consumption data from different communities allows cities to balance supply and demand across neighborhoods, optimizing distribution and reducing corresponding waste. Similarly, data on waste generation patterns, household recycling behaviors, and real-time bulk item pickups can be leveraged to optimize truck assignments and collection routes, promote circular practices and increase recycling rates and service efficiency as shown in Fig. 5.2. By breaking down data silos and encouraging

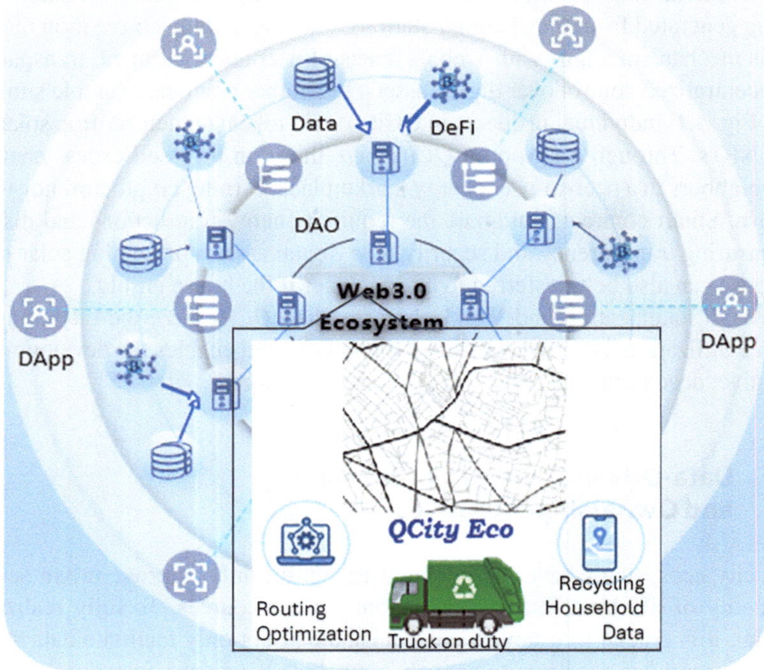

Fig. 5.2 Service improvements through data sharing

cross-sector integration, smart cities can create a dynamic ecosystem that not only enhances service delivery but also supports long-term urban resilience and environmental stewardship.

Incentivizing Citizen Data Contribution and Ownership

To encourage citizens to actively participate in data sharing, smart city ecosystems can implement incentive mechanisms that reward them for their contributions. One approach is to use a token-based system via DeFi within a Web3 framework. For example, citizens who share environmental data, such as air quality data from their home-based sensors or energy usage data from smart meters, can earn QCoins, NFT tokens described in Chap. 4.

These QCoins can then be used within the smart city ecosystem to access various services or benefits. Potential use cases include:

- Paying for public transportation or parking,
- Receiving discounts on energy-efficient appliances or environmentally friendly products,
- Accessing premium city services, and
- Participating in city governance decisions through voting or proposals.

By incentivizing data contribution, smart cities can create a more participatory and citizen-centric ecosystem, where individuals are not only data providers but also active stakeholders who benefit from improved services and develop a sense of ownership and pride in the city's development.

5.2.3 Stakeholders' Participation in Decision-Making in Smart City Ecosystems

Smart city ecosystems thrive on the active participation of diverse stakeholders in decision-making processes. This ensures that services are not only technologically advanced but also aligned with the actual needs and preferences of the communities they serve. Leveraged DAOs within the ecosystem provide a powerful mechanism to facilitate inclusive governance.

5.2.3.1 DAO-Driven Service Development and Improvement

DAOs can play a crucial role in the development and enhancement of urban services. Consider the example of urban waste management as shown in Fig. 5.3, a critical aspect of smart city sustainability.

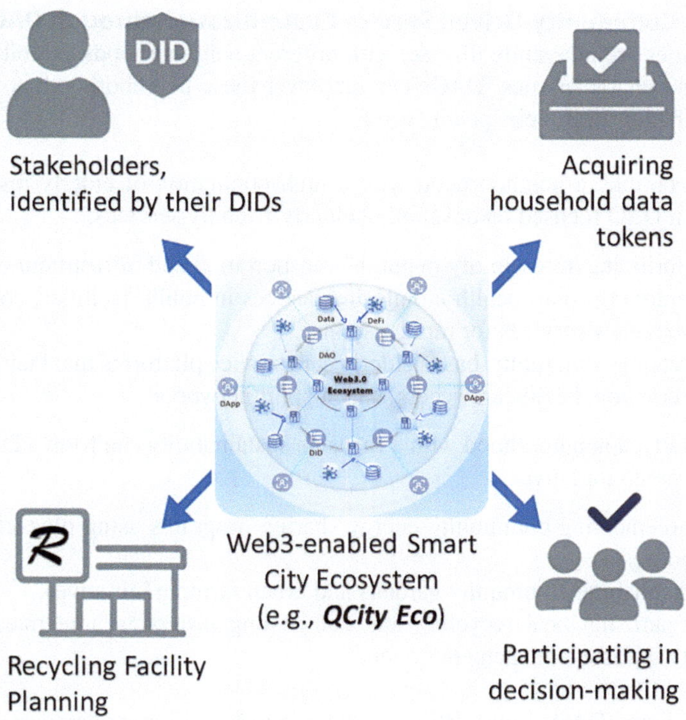

Fig. 5.3 Stakeholders' participation in recycling facility planning decision-making

- A DAO for urban waste management can be established, comprising members from various backgrounds:
  - Environmentalists with expertise in sustainable practices.
  - Waste treatment technicians with technical knowledge.
  - Logistics experts to optimize collection routes.
  - Data analysts to interpret waste generation patterns.
  - Ordinary citizens to represent diverse neighborhoods.
- This DAO can collaboratively develop innovative waste management solutions:
  - Designing user-friendly waste sorting systems with clear instructions and gamified incentives to increase citizen participation.
  - Utilizing IoT sensors and predictive analytics to analyze real-time waste generation data and plan dynamic, efficient waste collection routes.
  - Exploring and implementing advanced recycling technologies and circular economy models to minimize landfill use and maximize resource recovery.
- A specific example is the planning and construction of a new recycling facility. The decision-making process within *QCityEco* is transparent and implemented through DIDs, smart contracts, and DApps portals, ensuring accountability and trust among all stakeholders.

5.2.3.2 Community-Driven Service Customization through DAOs

Smart cities are inherently diverse, with different neighborhoods exhibiting unique needs and characteristics. DAOs can empower these neighborhoods to tailor services to better meet their specific needs.

- For example, a neighborhood with a high population of elderly residents can form a DAO focused on developing elderly-friendly services:
  - Prioritizing accessibility in public transportation and infrastructure.
  - Deploying smart health monitoring devices in public facilities, connected to emergency services for rapid response.
  - Creating community-based elderly care service platforms that facilitate social interaction, healthcare access, and support networks.
- Similarly, a neighborhood with a focus on sustainability can form a DAO to promote green initiatives:
  - Implementing community energy sharing programs using blockchain-based microgrids.
  - Establishing community gardens and urban farming initiatives.
  - Organizing local recycling and composting programs, integrated with the city's waste management system.

By utilizing DAOs, smart city ecosystems can foster a more democratic and participatory approach to urban governance. This approach ensures urban services are tailored to the needs of individual communities by giving citizens a direct voice in shaping their city's future.

5.3 Navigating Challenges in Web3-Driven Value Co-creation for the Smart City Ecosystem

The integration of Web3 technologies into smart city ecosystems holds immense promises for transforming urban living, offering opportunities to enhance services and empower citizens. However, realizing this potential requires careful consideration and proactive management of several inherent challenges. These challenges can be broadly categorized into technical hurdles that affect system functionality and social/regulatory issues that impact adoption and governance.

5.3.1 Technical Challenges: Ensuring Functionality and Scalability

One key technical challenge is interoperability. Smart cities are complex environments where various systems and data sources must work together. These systems may be built on different blockchain platforms or older infrastructure, which can create difficulties in communication and data exchange. For example, a decentralized energy grid management system built on Web3 might struggle to integrate with a city's existing traffic management system. Overcoming this requires developing standardized interfaces, cross-chain communication protocols, middleware solutions, and APIs to enable seamless interaction between different components of the Web3-empowered smart city ecosystem. Leveraging the open architecture approach can promote modularity, compatibility, and interoperability.

Another significant technical challenge is scalability. Smart cities generate vast amounts of data and involve numerous transactions and user interactions. As more participants and applications utilize Web3, the systems may face limitations in handling the increased load. For instance, a city-wide decentralized identity system or widespread use of DAOs for local governance could lead to network congestion, slow processing times, and higher computational costs. Addressing this involves adopting advanced blockchain technologies and architectural patterns, such as scalable consensus mechanisms, sharding techniques, layer-2 scaling solutions, off-chain data storage, and optimized data structures (Zamani et al., 2018; Jain et al., 2025).

5.3.2 Social and Regulatory Challenges: Fostering Adoption and Trust

Beyond the technical aspects, social and regulatory challenges also play a crucial role. Public awareness and education are paramount because Web3 technologies, such as NFTs, DAOs, and token-based economies, are still relatively new, and public understanding is limited. This lack of awareness can hinder citizen participation in Web3-driven initiatives. To address these challenges, cities must invest in comprehensive public education initiatives that encompass government-led awareness campaigns, accessible learning resources, hands-on workshops, and pilot programs. These efforts aim to demonstrate the benefits of Web3 while addressing common concerns around security, privacy, and usability.

Finally, regulatory uncertainty poses a major challenge. The legal frameworks for Web3 in smart cities are still evolving, creating ambiguity for businesses, developers, and governments. Issues such as the legal status of digital assets, data privacy, taxation, and DAO governance need clarification. Establishing a clear and supportive regulatory environment is crucial to fostering innovation and responsible development. This requires governments to actively participate in regulatory discussions, formulate appropriate local regulations, create regulatory sandboxes, and collaborate with stakeholders to develop guidelines that promote transparency, accountability, and ethical considerations.

5.4 Toward a Resilient, Equitable, and Sustainable Smart City Future

The smart city ecosystem is no longer a distant vision. It is an unfolding reality shaped by the interplay of technology, governance, and citizen engagement. As this chapter has demonstrated, resilience, equity, and sustainability are not separate ambitions but interdependent cornerstones and priorities that must be pursued together. Resilience demands cities that can anticipate, absorb, and adapt to shocks, whether environmental, economic, or social, without compromising their core functions. Equity ensures that the benefits of smart city innovations are accessible to all residents, bridging digital divides and fostering inclusive participation. Sustainability requires careful stewardship of resources, minimizing environmental impact while enhancing long-term sustainability and livability.

Web3 technologies offer an unprecedented opportunity to weave these priorities into the very fabric of urban governance and service delivery. By enabling decentralized decision-making, secure and self-sovereign identities, transparent transactions, and equitable value-sharing, Web3 empowers citizens to be co-creators, not just consumers, of urban innovation, thereby promoting and facilitating value co-creation. This chapter elaborates that DAOs, DID, DeFi, and DApps can transform assets, data, and services into shared community value, unlocking new models for participation and co-ownership.

Specifically, the Web3-empowered platform, *QCity Eco*, exemplifies this by allowing the tokenization of assets such as household solar energy generation and EV charging stations. Through this mechanism, the city promotes resource optimization and citizen participation, enabling decentralized services across various sectors. Decisions within the *QCity Eco* platform, such as planning a new recycling facility, are transparently implemented through DIDs, smart contracts, and DApps.

Yet, achieving this vision will require more than technical deployment. It calls for a cultural shift toward open collaboration, cross-sector integration, and a governance ethos rooted in trust, accountability, and inclusivity. Cities must invest in public education, cultivate interoperable systems, and develop adaptive policies that encourage experimentation while safeguarding public interest. The path forward is both challenging and inspiring. If pursued with foresight and collective commitment, the smart city of the future can become a living ecosystem, resilient against disruption, equitable in opportunity, and sustainable for generations to come. The decisions we make today will determine whether our cities remain fragmented infrastructures or evolve into vibrant, interconnected communities where technology serves humanity's highest aspirations.

References

Abraham, A., Zhang, Y., & Prasad, S. (2024). Smart city-wide intelligent emergency response system: An evacuation management framework utilizing next-gen communications and IoT. In *2024 IEEE 21st India Council International Conference (INDICON)* (pp. 1–7). IEEE.

Alam, T., Gupta, R., Nasurudeen Ahamed, N., Ullah, A., & Almaghthwi, A. (2024). Smart mobility adoption in sustainable smart cities to establish a growing ecosystem: Challenges and opportunities. *MRS Energy & Sustainability, 11*(2), 304–316.

Anthopoulos, L. G. (2017). The rise of the smart city. In *Understanding smart cities: A tool for smart government or an industrial trick?* (pp. 5–45). Springer.

Anthopoulos, L., & Fitsilis, P. (2010). From digital to ubiquitous cities: Defining a common architecture for urban development. In *The sixth international conference on intelligent environments* (pp. 301–306). IEEE.

Bressanelli, G., Adrodegari, F., Pigosso, D. C., & Parida, V. (2022). Towards the smart circular economy paradigm: A definition, conceptualization, and research agenda. *Sustainability, 14*(9), 4960.

Caputo, F., Magliocca, P., Canestrino, R., & Rescigno, E. (2023). Rethinking the role of technology for citizens' engagement and sustainable development in smart cities. *Sustainability, 15*(13), 10400.

Costa, D. G., Peixoto, J. P. J., Jesus, T. C., Portugal, P., Vasques, F., Rangel, E., & Peixoto, M. (2022). A survey of emergencies management systems in smart cities. *IEEE Access, 10*, 61843–61872.

Estevez, E., Cenci, K., Fillottrani, P., & Janowski, T. (2021). Review of international standards and policy guidelines for smart sustainable cities. In *Smart cities and smart governance: Towards the 22nd century sustainable city* (pp. 69–99). Springer.

Faliagka, E., Christopoulou, E., Ringas, D., Politi, T., Kostis, N., Leonardos, D., et al. (2024). Trends in digital twin framework architectures for smart cities: A case study in smart mobility. *Sensors, 24*(5), 1665.

Gupta, A., Badr, Y., Negahban, A., & Qiu, R. G. (2021). Energy-efficient heating control for smart buildings with deep reinforcement learning. *Journal of Building Engineering, 34*, 101739.

Humbal, A., Chaudhary, N., & Pathak, B. (2023). Urbanization trends, climate change, and environmental sustainability. In *Climate change and urban environment sustainability* (pp. 151–166). Springer Nature Singapore.

Ishida, T. (2002). Digital city Kyoto. *Communications of the ACM, 45*(7), 76–81.

Jain, A. K., Gupta, N., & Gupta, B. B. (2025). A survey on scalable consensus algorithms for blockchain technology. *Cyber Security and Applications, 3*, 100065.

James, P., Astoria, R., Castor, T., Hudspeth, C., Olstinske, D., & Ward, J. (2021). Smart cities: Fundamental concepts. In *Handbook of smart cities* (pp. 3–33). Springer International Publishing.

Jiao, Z., Ran, L., Liu, X., Zhang, Y., & Qiu, R. G. (2020). Integrating price-incentive and trip-selection policies to rebalance shared electric vehicles. *Service Science, 12*(4), 148–173.

Jittrapirom, P., Caiati, V., Feneri, A. M., Ebrahimigharehbaghi, S., Alonso-González, M. J., & Narayan, J. (2017). Mobility as a service: A critical review of definitions, assessments of schemes, and key challenges. *Urban Planning, 2*(2), 13–25.

Kociuba, D., Sagan, M., & Kociuba, W. (2023). Toward the smart city ecosystem model. *Energies, 16*(6), 2795.

Komninos, N., Kakderi, C., Mora, L., Panori, A., & Sefertzi, E. (2022). Towards high impact smart cities: A universal architecture based on connected intelligence spaces. *Journal of the Knowledge Economy, 13*(2), 1169–1197.

Martins, F., Patrão, C., Moura, P., & de Almeida, A. T. (2021). A review of energy modeling tools for energy efficiency in smart cities. *Smart Cities, 4*(4), 1420–1436.

Pandey, S. K., Khan, M. T., & Qiu, R. G. (2018, November). Harnessing big data and analytics solutions in support of smart city services. In *INFORMS International Conference on Service Science* (pp. 117–128). Springer.

Paskaleva, K., Evans, J., & Watson, K. (2021). Co-producing smart cities: A Quadruple Helix approach to assessment. *European Urban and Regional Studies, 28*(4), 395–412.

Pulkkinen, J., Jussila, J., Partanen, A., Trotskii, I., & Laiho, A. (2019). Smart mobility: Services, platforms and ecosystems. *Technology Innovation Management Review, 9*(9), 15.

Qiu, R. (2014). *Service science: The foundations of service engineering and management*. Wiley.

Qiu, R. (2024). Web3 and service science: Perfectly aligned to maximize the values of all stakeholders in the digital economy. *Digital Transformation and Society, 3*(4), 337–339.

Qiu, R., Zu, T., Qian, Y., Qiu, L., & Badr, Y. (2019). Leveraging big data platform technologies and analytics to enhance smart city mobility services. In *Handbook of service science, Volume II* (pp. 567–587). Springer.

Razmjoo, A., Gandomi, A. H., Pazhoohesh, M., Mirjalili, S., & Rezaei, M. (2022). The key role of clean energy and technology in smart cities development. *Energy Strategy Reviews, 44*, 100943.

Reba, M., Reitsma, F., & Seto, K. C. (2016). Spatializing 6,000 years of global urbanization from 3700 BC to AD 2000. *Scientific Data, 3*(1), 1–16.

Rong, Y., Xu, Z., Liu, J., Liu, H., Ding, J., Liu, X., et al. (2022). Du-bus: A realtime bus waiting time estimation system based on multi-source data. *IEEE Transactions on Intelligent Transportation Systems, 23*(12), 24524–24539.

Silva, B. N., Khan, M., & Han, K. (2018). Towards sustainable smart cities: A review of trends, architectures, components, and open challenges in smart cities. *Sustainable Cities and Society, 38*, 697–713.

Souri, A., Zarei, M., Hemmati, A., & Gao, M. (2024). A systematic literature review of vehicular connectivity and V2X communications: Technical aspects and new challenges. *International Journal of Communication Systems, 37*(10), e5780.

Szpilko, D., de la Torre Gallegos, A., Jimenez Naharro, F., Rzepka, A., & Remiszewska, A. (2023). Waste management in the smart city: Current practices and future directions. *Resources, 12*(10), 115.

Touloumidis, D., Xenou, E., & Ayfantopoulou, G. (2024). Empowering cities: Unveiling the innovative urban mobility ecosystem to support the transition through city-led innovations. In *Transport research arena conference* (pp. 31–45). Springer Nature Switzerland.

Troisi, O., Kashef, M., & Visvizi, A. (2022). Managing safety and security in the smart city: Covid-19, emergencies and smart surveillance. In *Managing smart cities: Sustainability and resilience through effective management* (pp. 73–88). Springer.

UN. (2018). *World urbanization prospects: The 2018 revision*. Retrieved January 10, 2025, from https://www.un-ilibrary.org/content/books/9789210043144

Zahra, A., Ghafoor, M., Munir, K., Ullah, A., & Ul Abideen, Z. (2024). Application of region-based video surveillance in smart cities using deep learning. *Multimedia Tools and Applications, 83*(5), 15313–15338.

Zamani, M., Movahedi, M., & Raykova, M. (2018). Rapidchain: Scaling blockchain via full sharding. In *Proceedings of the 2018 ACM SIGSAC conference on computer and communications security* (pp. 931–948).

Zhou, Y., & Liu, J. (2024). Advances in emerging digital technologies for energy efficiency and energy integration in smart cities. *Energy and Buildings, 315*, 114289.

Web3-Driven Value Co-creation in Future Smart Ecosystems

6

Abstract

As the global economy and public services undergo profound digital transformation, Web3 is emerging as a foundational decentralized infrastructure for holistic, inclusive, and ethically grounded value co-creation. In this chapter, we explore how Web3 diverges from Web2's centralized, platform-driven architecture by fostering a decentralized, interoperable, and human-centric ecosystem built on user self-autonomy and distributed governance. Within this paradigm, we articulate that users transition from passive data providers to value co-creators, empowered to shape and directly benefit from the digital services they actively engage with. As the final highlights of this book, by relying on the smart city context, particularly the conceptual QCity model, we examine three emerging technologies further driving Web3 innovation in the foreseeable future: (1) Intelligent Evolution through AI and Machine Learning (ML), which forms the cognitive core of decentralized systems; (2) Trusted Data Spaces (TDSs), which enable ethical, transparent, and privacy-preserving data collaboration through technologies such as Decentralized Identifiers (DIDs), smart contracts, zero-knowledge proofs (ZKPs), and Secure Multi-Party Computation (SMPC); and (3) the metaverse as a virtual frontier, a decentralized, AI-augmented immersive environment enabling participatory urban planning, virtual healthcare, and decentralized education. Ultimately, we argue that Web3-enabled value co-creation provides a holistic framework for smart, human-centric living, where technology, governance, and ethics align to redefine the future of human life.

Keywords

AI and machine learning · Trusted data spaces (TDSs) · Metaverse · Smart living · Human-centric living · Future smart system

R. Qiu, *Value Co-Creation in Web 3.0*, Advances in Data Analytics, AI, and Smart Systems, https://doi.org/10.1007/978-3-032-12911-6_6

6.1 Introduction

As cities, economies, and public services undergo profound digital transformation, Web3 is rapidly emerging as a foundational, technical infrastructure for systemic, inclusive, and ethically grounded value co-creation. Building on insights from the discussions in the preceding chapters, it becomes clear that this paradigm shift has marked a decisive departure from the centralized, frequently siloed, and organization-centric architecture that has characterized Web2. Instead, Web3 fosters a decentralized, interoperable, and human-centric ecosystem, defined by transparency, trust, user sovereignty, and distributed governance.

At its core, Web3 fundamentally reimagines the relationship between individuals, institutions, and intelligent systems. Users are no longer passive data providers; they become active co-creators, empowered to shape, control, and directly benefit from the services they engage with. This capacity for fluid collaboration among individuals, communities, organizations, and even artificial intelligence (AI) agents, or physical AI (NVIDIA, 2025), in the foreseeable future, all within trusted digital infrastructures, transforms not only how value is created but also who participates in its co-creation. This alignment of technological advancement with principles of inclusivity, resilience, and ethical design is no longer optional, but essential in our AI-augmented digital era (Qiu, 2024, 2025).

Web3's unique ability to dynamically orchestrate personalized, ethically engineered, and environmentally sustainable service interactions around people's daily life needs, at the very point of demand, makes it exceptionally well-suited for future smart ecosystems. Rather than treating users merely as data sources or consumers, once again, Web3 reconceptualizes them as active co-creators within context-aware, intelligent, and human-centric service networks.

As the final chapter of this book, it delves into three fast-changing yet interconnected data and computational technologies that underpin Web3's evolving potential to drive significant innovation and value co-creation, envisioning the rapid digital transformation that is surely shaping the future of smart living (Otto et al., 2022; Chen et al., 2023; Deshmukh et al., 2024; Hyland-Wood & Johnson, 2024; Cao et al., 2025; Wu et al., 2025; Zeng et al., 2025):

- Intelligent Evolution through AI and Machine Learning (ML): The integration of AI and ML within decentralized ecosystems enables the delivery of intelligent, adaptive, and personalized services, ranging from digital assistants to autonomous systems, including advanced humanoid robots. These technologies support decentralized decision-making and foster predictive, context-aware service delivery across all sectors, enhancing system capabilities, efficiency, and user experience.
- Trusted Data Spaces (TDS): As data becomes the lifeblood of digital services, TDS empowers individuals, organizations, and communities to retain granular control and ownership. Leveraging privacy-preserving technologies like zero-knowledge proofs (ZKPs) and governed by smart contracts or Decentralized Autonomous Organizations (DAOs), TDS ensures data is ethically shared, trans-

parently governed, and collaboratively used, ushering in a new model of trustworthy and equitable data exchange.

- Metaverse as a Virtual Frontier: Reimagined as a collective, decentralized virtual environment where physical and digital experiences converge, the future metaverse (supported by Web3 protocols, TDS, and AI/ML) offers unprecedented opportunities for value co-creation in education, governance, commerce, and culture development. It further and profoundly transforms how individuals, organizations, and communities interact, work, and co-create within hybrid realities.

Collectively and collaboratively, these rapidly evolving technologies demonstrate how Web3 will transform the future of digital interaction and service delivery from centralized control by moving into while empowering decentralized, intelligent, and ethically aligned ecosystems. The remaining sections of this chapter delve into each of these components in greater depth, examining how they empower all participants, whether human or artificial, to act as respected, intelligent, autonomous, and self-evolving agents in co-creating value within digitally enriched physical worlds for years to come.

6.2 Artificial Intelligence and Machine Learning in Web3: The Cognitive Core of Decentralized Value Co-creation

The integration of AI and ML within decentralized digital ecosystems marks one of the most transformative frontiers in the evolution of Web3. When combined with Web3's foundational pillars of distributed computing, blockchain-based governance, and privacy-preserving infrastructure discussed in Chap. 3, AI and ML form the cognitive engine of Web3-driven smart environments. These technologies transform Web3 from a data-centric infrastructure to a context-aware and intelligent value co-creation network, capable of delivering personalized, ethical, and autonomous services at scale and in real-time, at the point of demand.

Unlike traditional AI/ML models constrained by centralized architectures and data silos, AI/ML in a Web3 environment evolves into an autonomous and even self-evolving service actor, capable of making independent, context-aware decisions and participating meaningfully in decentralized systems. Leveraging agentic AI capabilities (Acharya et al., 2025), these intelligent agents are designed to perform goal-directed tasks, navigate complex multi-agent networks, and adapt dynamically to changing needs and environments, without relying on centralized control structures (Shen et al., 2024).

This form of distributed, proactive intelligence is especially critical within decentralized infrastructures such as DAOs. Here, participants and AI agents can co-manage operations, co-support consensus-building, co-optimize service delivery, and co-enforce policy decisions, all in accordance with governance protocols encoded on-chain in Web3. Their role extends beyond automation; they serve as collaborative stakeholders in the value co-creation process.

The following subsections explore how AI/ML capabilities within the Web3 ecosystem, particularly in areas such as data sovereignty, intelligent service delivery, participatory AI governance, and the emergence of physical AI, collectively enable and accelerate intelligent, ethical, and inclusive value co-creation.

6.2.1 Access to Provenance-Rich, Consent-Driven Data

A substantial limitation in the Web2 landscape is the "data-rich, information-poor" (DRIP) paradox (Qiu, 2024), where massive amounts of data are collected but often remain largely proprietary, inaccessible, unstructured, or ethically compromised. Web3 addresses this by embedding AI within trusted data networks, such as TDS, which are anchored in user consent, provenance verification, and decentralized access control. This ensures that AI/ML models are trained on high-quality, transparent, and community-sanctioned datasets, enhancing not only performance but also fairness, explainability, and accountability (Shen et al., 2024; Cao et al., 2025). This approach is crucial for smart city development, as discussed in Chap. 5, where integrating diverse electronic data sources into a unified, robust, trusted, and scalable platform is vital for effective city service management over time.

6.2.2 Intelligent, Context-Aware Services

AI/ML capabilities in Web3 enable highly adaptive, personalized services. AI/ML-powered platforms continuously learn from ambient environments, utilizing real-time data such as urban mobility patterns, energy consumption, or health biometrics, particularly within smart city contexts, as seen in QCity in Chap. 5. This capability facilitates the delivery of predictive, context-sensitive experiences (Shen et al., 2024). Applications, including decentralized applications (DApps), range from self-optimizing transportation grids that dynamically adjust traffic light timings to alleviate congestion, to autonomous energy distribution and responsive healthcare and decentralized emergency and safety management platforms. These services evolve in real time, delivering adaptive intelligence tailored to individual and community needs, as explored in detail in Chaps. 4 and 5. For instance, AI/ML and IoT can leverage archived historical data and real-time monitoring to enable predictive analytics for managing energy, traffic, safety, and public health in smart cities.

6.2.3 Toward Participatory AI/ML Governance

Web3 fundamentally reshapes AI/ML governance through decentralized and participatory frameworks, empowering users to actively engage in the oversight of algorithms. Smart contracts and DAOs allow communities to transparently vote on model behaviors, training datasets, updating cycles, and ethical constraints. This participatory AI/ML regulation ensures that AI/ML aligns with collective values and preserves accountability and privacy, transforming users from data subjects into co-governors of AI/ML (Shen et al., 2024; Kumari, 2025). This aligns with the smart city's need (e.g., QCity in Chap. 5) for collaborative governance and public-private partnerships that have reshaped urban planning and service delivery. AI/ML-augmented DAOs can play a crucial role in the development and enhancement of urban services, including carpooling and waste management, by facilitating collaborative design and transparent implementation through the use of smart contracts. They also empower communities to customize services, for example, by prioritizing accessibility for elderly residents or promoting customized green initiatives across cities.

6.2.4 AI/ML-Augmented Smart Contracts and Intelligent Services

AI/ML techniques, such as reinforcement learning and deep neural networks, can directly extend the functionality of smart contracts, transforming them from static rule executors into adaptive, self-optimizing service agents (Shen et al., 2024; Cao et al., 2025). For example, smart contracts can be augmented to dynamically adjust pricing in decentralized markets, proactively act on suspicious events, or reallocate computing resource demands to adapt to sudden surges in operational edge computing services. These AI/ML capabilities embedded in smart contracts can substantially facilitate the fast development of intelligent capabilities to empower various applications, such as decentralized finance (DeFi), digital supply chains, and public infrastructure systems, enhancing their operational efficiency and adaptability.

In a Web3-empowered smart city, physical assets, such as EV charging stations, underutilized vehicles, or rooftop solar panels (e.g., QCity in Chap. 5), can be tokenized as NFTs. AI/ML-augmented smart contracts then optimize transactions, enabling peer-to-peer energy marketplaces or decentralized service provision and exemplifying enhanced value co-creation for all participants.

6.2.5 Physical AI and Embodied Autonomy

The convergence of AI with physical systems, referred to as physical AI, enables embodied intelligence in real-world contexts (NVIDIA, 2025; Parada, 2025). In the Web3 ecosystem, drones, autonomous vehicles, Gemini Robotics, humanoid robots, and ambient sensors become intelligent, autonomous actors. Governed through smart contracts and trained via real-time feedback, these physical AI agents will participate in on-the-ground value co-creation, optimizing logistics, ensuring environmental compliance, and delivering emergency services at the point of need.

For example, in smart cities, IoT sensors and AI/ML, integrated as Physical AI components, are crucial for real-time monitoring of traffic, safety, and public health. Intelligent bus systems utilize GPS tracking and smart dispatching algorithms to make dynamic route adjustments, while V2V and V2I communication enhance traffic management and public safety. Smart sensors also play a foundational role in the early detection of hazardous events, triggering rapid alerts to emergency and safety management systems. At home, in addition to monitoring the safety of elderly individuals, a humanoid robot could help recreate a familiar, family-like atmosphere on a daily basis, supporting the maintenance or rejuvenating of cognitive function in older adults.

Summary: AI/ML as a Decentralized Value Catalyst

Ultimately, Web3 and AI/ML together lay the groundwork for the smart ecosystems of the future. As explored above, AI/ML technically forms the cognitive engine of Web3-empowered smart environments. In other words, Web3 delivers decentralization, data sovereignty, and trust, while AI/ML provides cognition, autonomy, and responsiveness. In conclusion, they collectively empower humans and machines alike to operate as autonomous, respected, and self-evolving agents in real-time, ethically, and inclusively, fostering value co-creation. This partnership not only redefines service delivery but also reimagines the very architecture of digital society, where intelligence is not centralized but democratized, co-governed, and self-evolving.

6.3 Trusted Data Spaces: Enabling Equitable and Secure Value Co-creation in Web3 Ecosystems

In the current digital economy, data has rapidly ascended to become a fundamental asset, powering innovation, intelligent decision-making, and economic competitiveness across organizations and societies. Data has been particularly considered as the lifeblood or new "oil" since the turn of the millennium. However, the exponential growth of data from various sources, spanning organizations, devices, and global borders, particularly in complex environments such as smart cities, has introduced critical challenges in its adequate and effective management, governance, fair and ethical usage, and value co-creation. To address these complexities, the concept of data spaces has emerged as a transformative solution (Solmaz et al., 2022). Within a defined data space, data sovereignty is a paramount objective, ensuring that data contributors retain granular control over the access, usage, and value exchange of their collected and archived data. A data space is often governed through sophisticated contractual and policy-based mechanisms, which are typically defined by data providers but are customarily tailored to the specific needs of data consumers when the stored data is delivered and consumed. At its core, a data space is an interoperable environment that enables secure collection, integration, sharing, and governance of data from diverse sources, while robustly respecting data ownership and the rights of providers within its defined context (Otto et al., 2022).

More recently, Trusted Data Spaces (TDS) represent the most advanced evolution of this concept (NTT, 2022). TDS is emerging as a socio-technical system that seamlessly blends robust data governance, stringent privacy protection, enhanced interoperability, and federated control. This sophisticated integration establishes unparalleled trust in both data and its manifold applications. By fundamentally transforming data into a shared and governed asset rather than a mere commodity to be extracted, TDS empowers organizations, individuals, and intelligent agents to collaborate securely and transparently (Schuiki et al., 2025). The prominence of TDS has accelerated significantly in recent years, propelled by global trends such as the pervasive integration of AI, the imperative for cross-border international data sharing, and the decentralization ethos of Web3 (Alsamhi et al., 2024; Wan et al., 2024; Zhang et al., 2025).

As the digital world continues its inexorable march toward decentralization (Ghosh & Gorai, 2024), TDS will become a pivotal enabler and upholder of Web3's core vision: fostering equitable, interoperable, and sovereignty-respecting ecosystems for data and value exchange. Unlike traditional platforms that centralize control and monetize user data, TDS is meticulously designed to empower data contributors, enforce ethical usage, and uphold accountability through well-defined, transparent governance frameworks. Crucially, they facilitate privacy-preserving collaboration, enabling AI systems and digital agents to operate effectively without compromising user autonomy or regulatory compliance while aligning well with the ethos of the Web3 era.

The evolution of TDS signals a broader, profound shift toward distributed, ethical, and value-driven data systems. Given that data is the new "oil," TDS, if incorporated, is definitely an indispensable component of the growing Web3 ecosystem. As AI and blockchain-based infrastructures continue to mature, TDS is poised to become the cornerstone of operations, trust, control, and collective intelligence in the Web3 era. Looking forward, sustained innovation in privacy-enhancing computation, AI-enabled governance, and cross-jurisdiction policy alignment will further extend the capabilities and global reach of these trusted data systems (Ghosh & Gorai, 2024; Cao et al., 2025).

The following subsections delve deeper into how TDS can be effectively adopted and implemented within the Web3 landscape. More specifically, their foundational attributes, architectural layers, interoperability mechanisms, application domains, and emerging trends are discussed, highlighting their indispensable role in shaping the future of secure, sovereign, and collaborative digital transformation.

6.3.1 Foundational Attributes of TDS in Web3

First and foremost, the effectiveness of TDS within the Web3 ecosystem hinges on a set of foundational attributes that collectively ensure data is collected and stored not merely as a raw resource, but as a sovereign, shared, and ethically governed digital asset in an easily accessible and readily usable manner (Otto et al., 2022). These attributes are crucial for enabling secure, accountable, and value-driven data sharing across decentralized environments, thereby reinforcing value co-creation, cross-board innovation, and mutual benefits among all participants in the Web3 ecosystem.

Granular Data Sovereignty

At the very heart of TDS lies the principle of granular data sovereignty, where individuals, institutions, and even intelligent agents as data providers retain explicit ownership and fine-grained control over their own data. By applying DIDs and smart contracts, these data providers can programmatically define precisely who can access their data, under what specific conditions, and for what explicit purposes (Zhang et al., 2025). For instance, leveraging blockchain technology, each transaction or data-sharing interaction within a TDS is inherently consensual, auditable, and legally enforceable, thereby ensuring robust accountability and regulatory compliance across various jurisdictions.

This approach is vividly exemplified in QCity, the prototype Web3-empowered smart city discussed in Chap. 5. When a TDS is applied, residents can manage their personal and sensor-generated data through an adopted decentralized identity framework. Through user-defined smart contracts, QCity citizens grant selective yet highly safeguarded access to urban service providers, such as transportation, energy, and healthcare organizations, based on real-time needs. The adopted TDS preserves user control while simultaneously enabling highly personalized and efficient services.

Privacy-Preserving Sharing Protocols

TDS integrates advanced privacy-enhancing technologies to facilitate secure data sharing and operational collaboration without compromising confidentiality. Cutting-edge techniques, including ZKPs, SMPC, and homomorphic encryption, allow for complex computation and analytics to be executed on encrypted or distributed data without ever revealing the underlying sensitive information (Knott et al., 2021).

In QCity, for instance, when a TDS is applied, public health data can be readily shared across institutions for epidemiological modeling using SMPC, enabling healthcare agencies to collaborate on critical research without exposing individual patient-level data. Similarly, traffic optimization modeling can effectively leverage encrypted citizens' mobility data to reduce congestion, while meticulously respecting commuter privacy and significantly enhancing service efficiency.

Decentralized, Community-Governed Access and Compensation

Governance within a TDS is inherently decentralized, executed primarily through DAOs within the Web3 ecosystem. These community-led frameworks empower stakeholders to collaboratively define the rules for data access, usage permissions, pricing models, and equitable compensation. In QCity, neighborhood-based DAOs govern local data spaces, ensuring that data contributors, whether individuals, businesses, or AI agents, are fairly compensated for their contributions to various public services from civic applications to digital infrastructure planning. This community-first governance fosters unparalleled transparency, autonomy, and collective responsibility, transforming data into a co-managed, privacy-preserving public asset rather than a commodified resource fully controlled by centralized platforms in Web2.

6.3.2 System Architecture and Interoperability

The architecture of TDS within the Web3 ecosystem can be meticulously designed to ensure paramount security, scalability, and cross-domain interoperability. This robust design supports ethical and sovereign data exchange within today and tomorrow's complex digital environments. Essentially, TDS operates as a layered, modular system, facilitating flexible integration across diverse organizations, platforms, and national boundaries, a critical requirement for realizing dynamic and truly trusted data ecosystems (NTT, 2022; Zhang et al., 2025).

Authentication and Authorization Modules

At the foundational level, TDS implements sophisticated DID and access control mechanisms. These modules enable the secure verification of participants through decentralized credentials, such as verifiable credentials (VCs) and DIDs themselves. This architecture enables granular, role-based access management, ensuring that data is accessible only to authorized agents under precisely defined conditions.

In QCity, for example, citizens utilize self-sovereign identities to securely access a wide range of city services, such as healthcare and urban mobility. These individual digital identities are stored on-chain, empowering individuals to manage their own credentials and dynamically authorize defined and limited access rights for service providers or AI agents. This approach ensures that only necessary personal information is disclosed and utilized, bolstering data privacy and security.

Audit Trails and Traceability Engines

To cultivate and maintain trust while ensuring accountability, TDS incorporates immutable audit trails that meticulously log every data interaction, from initial request and access to deletion and derivation. These comprehensive logs are typically secured using blockchain technologies, which inherently ensure tamper-resistance and provide verifiable provenance tracking across the entire data lifecycle.

In QCity, comprehensive audit logs are maintained for all interactions involving public and private data. For instance, when environmental sensors capture air quality or vehicle CO_2 emissions, each subsequent use, whether by healthcare researchers or governmental policymakers, is immutably recorded, enabling full traceability and streamlined compliance reporting.

Data Lifecycle and Federation Management

Modern TDS supports a complete data lifecycle model, encompassing data creation, collection, validation, storage, sharing, reuse, and secure destruction. More importantly, TDS promotes and enables federated data space management, a paradigm in which multiple independent yet interconnected TDSs form a cohesive data federation (NTT, 2022). This fundamentally shifts away from isolated data silos toward distributed and interoperable networks, aligning well with a core principle of Web3.

In QCity, multiple domain-specific or local data spaces can be developed, covering vital city service sectors such as transportation, energy, healthcare, and the environment. They operate as autonomous, federated nodes and synchronize seamlessly with one another and with external partners through a Web3-empowered orchestration layer, allowing for real-time cross-sector collaboration while maintaining meticulous domain-specific governance.

Interoperability Gateways

TDS fosters true interoperability through standardized connectors that seamlessly bridge disparate systems and platforms while ensuring strict regulatory compliance. Tools like International Data Spaces (IDS) connectors, Data Access Policy Services (DAPS), and advanced ontology alignment mechanisms facilitate smooth, cross-border, and cross-industry data flows (NTT, 2022). This enables secure data exchange even across jurisdictions governed by distinct privacy laws (e.g., GDPR, CCPA) (Almashaqbeh & Solomon, 2022; Zhang et al., 2025).

When fully digitalized, QCity can concretely demonstrate this by employing robust interoperability gateways to exchange data with national and international networks, such as cross-municipal transportation systems, global energy grids, or homeland security networks. For example, this capability can enable AI agents within QCity to coordinate effectively with external agents in critical areas, such as carbon tracking, disaster response, supply chain optimization, national security, and international trade, all while significantly improving collaboration and maintaining legal and ethical compliance.

In summary, this well-crafted and sophisticated architectural model will ensure that TDSs transcend their role as mere data storage or exchange platforms, becoming powerful enablers of sovereign, secure, and interoperable constituent digital systems, indispensable to Web3. As QCity vividly demonstrates, by embedding these architectural principles and utilities into the very fabric of urban operations, Web3-empowered cities can overcome various challenges explored in Chap. 5 to enable unprecedented living services for promising urban daily lives, achieving scalable trust, responsive governance, and ethical intelligence at a population scale.

6.3.3 Applications in the Web3 Ecosystem

TDS is fundamentally redefining how data is governed, exchanged, and utilized across boundaries, serving as critical infrastructure for the next generation of decentralized service system models. Within the Web3 ecosystem, TDS empowers participants to engage in secure, sovereign, and value-aligned data sharing and collaboration, enabling innovations that would be infeasible under traditional, centralized data architectures. The following highlights some compelling examples of how TDS can be well adopted and incorporated into the Web3 smart city ecosystem.

Smart Cities and Human-Centric Services

TDS is central to the design of smart cities that are truly responsive, equitable, and ethically intelligent. By leveraging these trusted data environments, citizens can securely share mobility, environmental, and behavioral data to access highly

AI-personalized public services, from dynamic transport routing to adaptive energy use, all while retaining full and granular control over their data.

In QCity, for instance, residents actively opt into real-time data exchanges with urban systems via decentralized identity wallets. In return, they receive context-aware benefits such as predictive healthcare alerts, automated waste collection based on real-time fill levels, public transportation information, or dynamic utility pricing, all managed through transparent smart contracts governed by local DAOs and TDSs. This exemplifies a symbiotic relationship between citizens and urban infrastructure, built on trust and mutual value.

Sustainable and Transparent Supply Chains

Federated TDSs are transforming industrial collaboration, particularly in the critical context of sustainability and environmental, social, and governance (ESG) performance. Initiatives like Catena-X exemplify how TDS enable automotive manufacturers and suppliers to exchange traceable data on carbon emissions, sourcing practices, and compliance metrics, ensuring verifiable ESG performance across complex supply networks (Zhang et al., 2025).

QCity's procurement system for road infrastructure management can be strategically integrated with similar TDS-enabled supply networks. This allows the municipality to source construction materials or energy services with full lifecycle transparency, thereby aligning urban development directly with ambitious circular economic goals and verifiable sustainable practices.

Decentralized and Ethical Healthcare Ecosystems

TDS is advancing the development of privacy-preserving, patient-centered healthcare models. By enabling the storage of health records in decentralized data vaults, patients retain unprecedented sovereignty over their medical information. They can then selectively share anonymized data for vital research, clinical trials, or personalized treatment plans, with cryptographic guarantees of privacy and control.

In QCity, citizens utilize their TDS-linked health profiles to receive proactive, preventive health recommendations from AI-driven, evolving agents, connect securely with remote care providers, or ethically donate anonymized data to local research institutions, all within a robust, consent-based, and transparent governance framework.

Cross-Border Data Collaboration

Beyond sectoral applications, TDS also facilitates secure, policy-compliant data exchange across national and organizational boundaries. Interoperability frameworks, such as those promoted by the NTT Group, the Smart Connected Supplier Network (SCSN), and the International Data Spaces (IDS), ensure that local compliance (e.g., GDPR, Act on the Protection of Personal Information) is meticulously maintained while enabling global innovation and collaboration (NTT, 2022).

QCity can actively participate in a cross-border climate alliance that leverages federated TDSs to exchange environmental impact data with peer cities across continents. This collaboration enables sophisticated AI-based risk modeling, policy benchmarking, and joint climate adaptation initiatives, all while preserving the privacy and regulatory adherence of raw, sensitive underlying data.

In summary, across these diverse applications, TDSs enable data to become a trusted, governable asset and utility, not only powering intelligent data sharing and collaboration but also fundamentally reinforcing digital sovereignty, ethical governance, and equitable value distribution. As robustly demonstrated in QCity and proposed in Fig. 3.3 of Chap. 3, the integration of TDSs within a Web3 framework will undoubtedly create the essential foundation for participatory, adaptive, trustful, and truly data-driven urban futures.

6.3.4 Future Prospects and Integration with AI and Web3

The continued maturation of TDSs represents far more than a mere safeguard for data privacy; it constitutes a cornerstone for the development of trustworthy, autonomous intelligence within Web3-empowered ecosystems. As decentralized systems increasingly rely on AI agents to drive intelligent services, the quality, provenance, and consent behind the data they consume become absolutely mission-critical. TDSs are uniquely positioned to fulfill this role by embedding robust governance, accountability, and ethical design principles directly into the data pipeline itself. In doing so, they enable AI and Web3 to evolve symbiotically toward ecosystems that are not only intelligent but also inherently just, transparent, and participatory.

TDSs support the future of AI and Web3 in at least the following three critical ways:

- Ethical, Explainable AI: AI models trained within TDS frameworks benefit from clean, verifiable, and consent-driven datasets. This ensures their outputs are fair, transparent, and accurately aligned with human intent, significantly reducing the risk of algorithmic bias and fostering greater public trust in AI-powered services.
- Agentic and Privacy-Preserving Intelligence: TDSs enable agentic AI systems to make context-aware, autonomous decisions without directly accessing raw or sensitive data. Through advanced privacy-preserving computation (e.g., federated learning, ZKPs), these agents can operate effectively in highly regulated environments, such as finance, healthcare, or governance, without compromising personal or institutional confidentiality.
- Tokenized Data Monetization and Fair Compensation: Within Web3 ecosystems, TDSs facilitate the development of innovative, decentralized data economies where individuals and organizations are fairly compensated for the value their data generates. Smart contracts automate this process, supporting new business models based on Data-as-a-Service (DaaS) and true value co-creation, moving beyond traditional data extraction models.

In QCity illustrated in Chap. 5, these transformative principles are already being explored through intelligent services that compensate citizens for sharing mobility or health data, utilize TDSs to train municipal AI agents, and implement community-voted ethical guidelines for AI behavior. This ambitious vision aligns with broader global initiatives such as Gaia-X, Catena-X, and the Data Society Alliance, all of which are converging on TDS as a critical building block of a global digital commons (NTT, 2022).

Summary: TDS: The Cornerstone of Web3's Ethical Data Ecosystem
TDS represents a foundational shift in how data is governed and valued, serving as the ethical infrastructure of the Web3 era. By striking a balance between innovation and individual sovereignty, TDS enables secure, equitable, and accountable data-driven progress. As socio-technical systems, TDSs embed trust directly into smart ecosystems, reconciling the needs of AI autonomy, decentralized governance, and data protection. Demonstrated with their great potential in global initiatives, TDSs are poised to become the indispensable lifeblood supply system of trusted digital infrastructures, ensuring that as we build smarter, AI-augmented service systems, we also build fairer and more human-centered societies in the Web3 era.

6.4 The Metaverse and Augmented/Virtual Reality: Immersive Layers of the Value Ecosystem

In the rapidly evolving digital ecosystem landscape, the metaverse and Augmented/Virtual Reality (AR/VR) are emerging as creative, enticing, and powerful immersive dimensions for value co-creation in Web3 environments. These fast-growing technologies transcend mere entertainment or emulation, representing interactive, persistent, decentralized, and immersive digital spaces where identity, agency, creativity, and value converge to redefine how individuals, organizations, and intelligent agents interact and co-grow (Ghosh et al., 2024).

At the heart of this immersive transformation, Web3, as defined in this book, provides the fundamental infrastructure for decentralization, trust, ownership, and value co-creation. When integrated with blockchain, self-sovereign identity (SSI), and privacy-preserving cryptographic frameworks within Web3, the metaverse becomes more than just a virtual reflection of the physical world; it transforms into a programmable, immersive, and participatory arena where all participants trustfully and collaboratively co-create value in real time. As Gebre et al. (2024) emphasize, these immersive environments can be made inherently secure and interoperable through blockchain-based architectures and DID frameworks, as shown in Fig. 3.3 in Chap. 3, enabling seamless, privacy-respecting, and enjoyable digital experiences across diverse platforms.

Simultaneously, social research in brand-based metaverse communities, such as the Gucci Discord case study (Sron & Heitmayer, 2024), reveals a growing trend of Community-Integrated Brand Engagement (CIBE). Here, end-users are no longer passive consumers but active contributors, co-authoring symbolic narratives, generating digital art, and emotionally investing in shared digital utopias. This grassroots creativity, identity construction, and mutual empowerment mark a new kind of psychologically meaningful digital citizenship and ownership, where emotional safety, self-expression, and a sense of belonging can thrive.

Thus, the metaverse, anchored by AR/VR and governed through Web3 protocols, can serve as a next-generation digital service space. It enables collaborative urban planning, decentralized education, virtual healthcare delivery, gamified sustainability campaigns, and other innovative applications. These integrated environments are context-aware, agent-enabled, and increasingly sovereign, blurring the boundaries between virtual interaction and tangible, real-world outcomes. The subsections that follow will explore how these further enhanced immersive virtual layers extend the traditional service ecosystem into embodied, co-creative domains, where smart, *deepservice* experiences can be cultivated, personalized, ethically aligned, and socially enriching (Qiu, 2024).

6.4.1 Decentralized Ownership and Participatory Immersive Environments

One of the most transformative features of the Web3-empowered metaverse is its architecture of decentralized ownership and participatory governance. In stark contrast to traditional, platform-centric virtual worlds, where corporations retain centralized control over access, monetization, and user data, Web3-empowered metaverse environments offer community-driven and participatory immersive alternatives firmly grounded in blockchain-based identity, asset ownership, and DAO decision-making.

By fully leveraging the tools and utilities enabled by the building blocks of Web3, such as DAOs, DIDs, DeFi, and TDSs, participants in these Web3-empowered spaces can acquire direct and verifiable ownership of virtual land, avatars, digital goods, and other available services. More importantly, they are usually motivated and encouraged to participate directly in the shaping of metaverse ecosystems and their future. Thus, all participants collectively and collaboratively define their common needs, governance policies, and community norms and values. This participatory model essentially facilitates the development of fairer, more transparent, and more inclusive business and societal service systems. As a result, the Web3 metaverse has shifted from a classic, corporate-controlled emulation to a vibrant, civic digital commons that supports better innovation and expression.

As discussed earlier, the principles of TDS, including data sovereignty, privacy-preserving sharing, and decentralized governance, can provide the needed data infrastructure for these immersive environments. Just as TDSs allow citizens in QCity to retain control over their personal and sensor data while being well engaged in real-time, AI-enhanced public services, the metaverse extends these rights into the experiential and living domain. Here, participants are not only data contributors but also creative co-owners of the digital environments they inhabit and help build. This fusion of decentralized technologies and immersive interfaces paves the way for new forms of urban engagement, education, commerce, and living. For example, virtual town halls, community-driven art districts, tokenized learning hubs, and interoperable service layers allow both human and artificial agents to collaborate fluidly across virtual and physical realms, aligning with the grand future and broader vision of smart *deepservice* that are heavily explored and strongly proposed throughout this book.

6.4.2 AI-Driven Personalization and Digital Twins

Recently, the seamless convergence of AI/ML and digital twin technologies has fundamentally redefined interactivity within immersive Web3 environments (Bamakan & Far, 2025). These intelligent systems, powered by real-time, provenance-rich data from reliable sources such as TDSs, enable the metaverse and AR/VR layers to be applied in various fields, delivering deeply personalized, context-aware, and continuously adaptive experiences across multiple sectors in daily life.

For example, AI-powered avatars and digital agents embedded within these environments can learn from dynamic user inputs, including health indicators, mobility patterns, environmental context, and even emotional cues. These resulting agents function not only as responsive interfaces but as autonomous, even self-evolving service actors. As conceptualized in Chap. 5, an individual can co-create value through well-engaged participation and real-time decision-making. In QCity, for instance, the integration of AI with federated TDS infrastructure can allow digital twins to mirror real-world infrastructure systems, such as public transit networks, energy grids, and public health dynamics, using high-fidelity, interactive simulations or emulations. These serve as intelligent proxies for urban experimentation, civic engagement, and the optimization of public services.

In practice, in addition to the manufacturing application in industry (Bamakan & Far, 2025), the immediate key application domains can also start with the following:

- Education: AI-driven tutors within immersive virtual classrooms dynamically adapt instructional content to individual learning styles, pacing, and pedagogical engagement, significantly enhancing both accessibility and learning outcomes.
- Healthcare: Personalized digital health twins emulate physiological responses to potential treatments, allowing doctors and patients to explore therapeutic options in a risk-free, personalized, and data-informed environment.
- Urban Planning: AI agents embedded within metaverse-based urban simulations guide users, such as planners, citizens, and policymakers, through participatory design processes. By referencing live city data from TDSs, they help visualize the impacts of planned infrastructure upgrades, conduct scenario planning, and inform democratic decision-making to meet the needs of most participants.

Once again, these intelligent services align with the broader concept of smart *deepservice*: services that are not only intelligent and personalized but also ethically grounded, context-aware, and co-produced in real time (Qiu, 2024). In the immersive layers of the Web3 smart city ecosystem, the fusion of AI cognition and digital embodiment thus becomes a powerful tool for responsive, democratic, and human-centered innovation and value co-creation.

6.4.3 Spatial and Experiential Value Co-creation

In Web3-empowered immersive environments, value co-creation is no longer confined to digital transactions or static interactions. It becomes spatially embodied, socially constructed, and experientially enriched. Through the convergence of AI agents, DAOs, DIDs, and spatial computing, participants and intelligent digital agents can collaborate in shared virtual environments that effectively mirror and augment real-world contexts, quickly expanding the potential of Web3-empowered metaverses.

Indeed, this immersive co-creation model can now empower users to participate in activities that were once the exclusive domain of institutional actors, such as policy design, urban development, or environmental planning. For instance, residents in a Web3-empowered smart city like QCity can enter a virtual replica of their neighborhood to propose changes to public space layouts, explore infrastructure scenarios through digital twins, or debate zoning decisions in real time within DAO-governed virtual town halls. These metaverse-powered spaces enable policy simulations, interactive education, and civic deliberation, all grounded in real-time data flows from TDSs. Such platforms democratize access to expertise by providing intuitive, visual, and interactive experiences that significantly reduce technical and cognitive barriers to participation. This aligns with the participatory principles detailed in Chap. 3, where TDS-backed services enable collaborative governance and adaptive service design, now seamlessly extended into high-dimensional spatial environments.

In addition to civic engagement, Web3-empowered metaverses governed by DAOs in QCity can also serve as living laboratories for the following:

- Social experimentation: Exploring alternative economic systems, social contracts, or collective behaviors
- Climate action planning: Simulating the environmental impact of proposed regulations, infrastructure changes, or community-led initiatives
- Cultural heritage preservation: Enabling immersive storytelling and memory-sharing that reconnect communities with their identities and shared histories

In the above-explored environments, co-creation becomes performative: people do not just consume information. In fact, they embody roles, test consequences, and shape outcomes alongside algorithms, digital twins, and fellow participants. As envisioned in QCity and supported by the ethical data architecture of TDS, this spatial and experiential layer of value co-creation represents a critical shift, from user engagement to citizen agency, from passive digital consumption to active socio-technical participation.

6.4.4 Cross-Reality Continuity and Socio-Technical Feedback Loops

A defining innovation in the convergence of Web3 and extended reality (XR) technologies lies in their ability to enable cross-reality continuity, a seamless, bidirectional interface between the digital and physical worlds (Thaker & Heitmayer, 2025). In this model, virtual interactions are no longer isolated simulations; they become operational components of real-world service ecosystems, directly interacting, both passively connecting and actively responding to, physical outcomes.

Enabled by AR/VR overlays, AI agents, and real-time data flows from TDSs, this synchronization establishes continuous socio-technical feedback loops. These loops connect humans, autonomous systems, and physical infrastructure, allowing cities to become not just responsive, but reflexive while learning from each interaction and dynamically adapting services to meet evolving needs. In QCity, this is already being explored in Chap. 5. Residents who participate in a virtual urban planning session can immediately trigger policy recommendations that are fed into smart contracts for further deliberation and DAO-based voting. Meanwhile, digital reputations built on blockchain, via participatory engagements, data contributions, or AI-assisted co-creation, can influence access to both digital and physical services, ensuring optimal alignment between community contribution and resource allocation.

In QCity, some immediate and illustrative scenarios can start with the following:

- Policy Co-design to Implementation: A zoning simulation run in the metaverse provides real-time feedback to city planning dashboards, which in turn update live municipal frameworks and regulations, which can be executed through real-time, community-driven voting.
- Augmented Maintenance with AI & IoT: AR/VR interfaces guide participants through public infrastructure repairs (e.g., bike racks, building escalators/elevators, smart lighting), informed by AI diagnostics from embedded IoT sensors and secured via blockchain-stamped service logs. Participants can voluntarily test and monitor the outcomes virtually.
- Cross-Reality Social Reputation: A blockchain-based reputation system that reflects one's contributions in both virtual community forums and real-world volunteering (e.g., sustainability projects or civic hackathons) may influence personalized service eligibility, incentives, or voting rights within a DAO-governed local network.

Notably, these feedback mechanisms are both technically viable and governance-aware, enabled by TDSs that uphold data provenance, user consent, and ethical accountability and supported by DAOs. This reinforces a new paradigm where virtual actions are not abstracted from reality but are deeply entangled with it, giving rise to co-evolving systems of trust, service delivery, and collective intelligence. Ultimately, cross-reality continuity transforms the smart city into a living socio-technical ecosystem, capable of sensing, learning, and evolving alongside its citizens (Liu et al., 2024). It exemplifies the full potential of Web3 not merely as infrastructure, but as an experiential fabric that integrates identity, agency, and ethical interaction across service sectors.

Conclusion: Co-creating Reality: Immersive Technologies in the Web3 Ecosystem

As metaverse and AR/VR technologies mature and become increasingly integrated into daily life, Web3 provides an ethical, technical, and governance infrastructure necessary to transform immersive environments into civic, economic, and ecological spaces of collective intelligence and co-creation. These environments are no longer limited to entertainment or simulation; they have evolved into an integral part of inclusive, participatory service ecosystems where people and intelligent agents co-create value in real time.

Evidently, the core of this transformation is driven by the convergence of DAOs, DIDs, DeFi, TDSs, AI-powered personalization, and edge and spatial computing. Together, they enable the rise of a metaverse that is ethical by design, interoperable by architecture, and co-owned and co-evolving by its participants. Users are not merely navigating virtual worlds; they are actively engaging with them in real time. Moreover, they are actively shaping them, building equitable economies, and reimagining governance, culture, and sustainability through immersive engagement.

As illustrated in QCity and grounded in the principles outlined in Chap. 5, this metaverse layer becomes a critical extension of the smart city ecosystem, where cross-reality continuity connects virtual actions with physical consequences and socio-technical feedback loops reinforce transparency, adaptability, and shared responsibility. With this grand vision, the metaverse and AR/VR/XR are no longer just a digital escape or overlay; they are a foundational work and living space for democratic innovation, human-centered design, and multiagent collaboration. It is where the future of public services, education, infrastructure, and identity can be collectively prototyped and continuously co-redefined and co-evolving, intelligently, inclusively, and sustainably.

6.5 Conclusion: Value Co-creation in Web 3.0 Through Blockchain, Metaverse, and Privacy-Preserving

As we approach the era of Web3, the very foundation of how value is created, exchanged, and shared is fundamentally rewoven on decentralized networks. This ongoing digital transformation will liberate us from the confines of centralized platforms and traditional transactional economies. As comprehensively explored in this book, we understand that value co-creation and decentralized distribution are poised to become the new standard. Our future business and society must be developed by relying on a dynamic, participatory, and ethically grounded model. Fundamentally, this shift is powered by the seamless integration of blockchain-enabled trust, immersive metaverse experiences, and privacy-preserving intelligence.

In this transformative paradigm, as articulated in this book, blockchain transcends its role as a mere ledger of transactions, establishing itself as a foundational infrastructure for transparency, provenance, and decentralized governance. As illustrated in Fig. 3.3 in Chap. 3, smart contracts, DAOs, DIDs, DeFi, and TDSs then collectively offer various interoperable and programmable Web3 functional utilities. As a result, the reliance on centralized intermediaries on a decentralized network becomes obsolete.

The widespread development of this Web3 framework and architecture truly marks the emergence of the Web3 era. We believe that Web3 will meet the challenging needs of fast-growing digital economies, promisingly promoting the collaborative creation and sharing of value in digital assets in real time through the empowerment of all stakeholders, including individual participants, organizations, governments, and intelligent AI agents.

Concurrently, the metaverse adds an essential embodied and experiential layer to this value co-creation. Immersive AR/VR/XR environments, fueled by real-time data and personalized AI agents, enable co-design, co-learning, and co-governance at a human scale. Whether simulating climate policies, reimagining urban spaces, or co-building virtual institutions, these environments transform value co-creation from an abstract concept into a concrete, living experience.

At the heart of this comprehensive transformation lies a growing imperative: data sovereignty and ethical intelligence. Web3's unwavering commitment to privacy-preserving technologies, such as ZKPs, homomorphic encryption, and SMPC, ensures that data is not exploited but rigorously protected and respected. When TDS is fully incorporated, individuals confidently maintain control over their digital identities and precisely determine how their data contributes to collective intelligence, thereby unlocking collaborative innovation without compromising personal autonomy.

Collectively, all these foundational technologies converge in what Qiu (2024, 2025) terms smart *deepservice*. The next generation of Web3-driven intelligent services can be meticulously personalized, human-centered, and ethically designed. The enabled services can be context-aware and self-evolving. Built on Web3 principles, future smart *deepservice* ecosystems will be more than interoperable and resilient. They will be designed for decentralized governance, data sovereignty, and user empowerment, fostering a new era of ethical and participatory value co-creation. These ecosystems will also be self-evolving, continually fueled by dynamic feedback from users, autonomous agents, and their environments.

In a visionary smart ecosystem like QCity, smart *deepservice* can be tangibly realized and continue to evolve through a comprehensive, interplayed-layer architecture:

- Open, data-driven service markets that enable real-time value exchange between AI agents, citizens, and institutions
- Tokenized data product registries and exchanges that facilitate verifiable ownership and monetization of data assets

- Privacy-first computing environments that preserve trust, integrity, and compliance across diverse sectors and international borders
- Governance and participation layers that empower stakeholders to collectively manage the ecosystem through decentralized decision-making
- Digital identity and reputation layers that facilitate trust and access to services without compromising personal data
- Physical-digital integration layers that connect the urban environment to the Web3 ecosystem via a pervasive network of IoT sensors and digital twins

What ultimately emerges is not simply a decentralized digital infrastructure, but a new societal operational and living environment. It is truly one where value is co-created by co-design and participation, agency is distributed equitably, and digital experiences are deeply rooted in ethical, inclusive, and adaptive intelligence. Fundamentally, Web3 redefines the value of data assets, not as a commodity to be extracted, but as a highly rewarding outcome of genuine collaboration, collective co-governance, and continuous co-evolution. Blockchain ensures trust. The metaverse makes it experiential. Privacy preservation makes it ethical. Promisingly, the resulting smart deepservice makes it profoundly human.

The technology enabling value co-creation in Web3 is not merely the future of technology; it is the definitive future of smart living.

References

Acharya, D. B., Kuppan, K., & Divya, B. (2025). Agentic AI: Autonomous intelligence for complex goals—A comprehensive survey. *IEEE Access, 13*, 18912–18936.

Almashaqbeh, G., & Solomon, R. (2022, June). Sok: Privacy-preserving computing in the blockchain era. In *2022 IEEE 7th European Symposium on Security and Privacy (EuroS&P)* (pp. 124–139). IEEE.

Alsamhi, S. H., Hawbani, A., Kumar, S., Timilsina, M., Al-Qatf, M., Haque, R., et al. (2024). Empowering dataspace 4.0: Unveiling promise of decentralized data-sharing. *IEEE Access, 12*, 112637.

Bamakan, S. M. H., & Far, S. B. (2025). Distributed and trustworthy digital twin platform based on blockchain and Web3 technologies. *Cyber Security and Applications, 3*, 100064.

Cao, B., Xiao, S., Shi, L., Wang, T., Chen, J., Wang, J., et al. (2025). Web 3.0: A survey on the architectures, enabling technologies, applications, and challenges. *IEEE Communications Surveys and Tutorials*.

Chen, H., Duan, H., Abdallah, M., Zhu, Y., Wen, Y., Saddik, A. E., & Cai, W. (2023). Web3 Metaverse: State-of-the-art and vision. *ACM Transactions on Multimedia Computing, Communications and Applications, 20*(4), 1–42.

Deshmukh, R. A., Collarana, D., Gelhaar, J., Theissen-Lipp, J., Lange, C., Arnold, B. T., Curry, E., & Decker, S. (2024). Challenges and opportunities for enabling the next generation of cross-domain dataspaces. In *The Second International Workshop on Semantics in Dataspaces, Co-located with the Extended Semantic Web Conference*.

Gebre, D., Hadish, S., Sbhatu, A., Aloqaily, M., & Guizani, M. (2024). Establishing trust and security in decentralized metaverse: A Web 3.0 approach. *ACM Transactions on Multimedia Computing, Communications and Applications, 20*(12), 1–17.

Ghosh, S., & Gorai, S. (2024). *The age of decentralization: How Web3 and related technologies will change industries and our lives*. Productivity Press.

Ghosh, A., Hassija, V., Chamola, V., & El Saddik, A. (2024). A survey on decentralized Metaverse using Blockchain and Web 3.0 technologies, applications, and more. *IEEE Access*.

Hyland-Wood, D., & Johnson, S. (2024). Intersections of Web3 and AI—View in 2024. *arXiv preprint arXiv:2411.04318*.

Knott, B., Venkataraman, S., Hannun, A., Sengupta, S., Ibrahim, M., & van der Maaten, L. (2021). Crypten: Secure multi-party computation meets machine learning. *Advances in Neural Information Processing Systems, 34*, 4961–4973.

Kumari, P., (2025). *Modern AI technologies - Generative AI, agentic AI, and AI agents*. Data Science & Analytics. Retrieved July 25, 2025, from https://www.linkedin.com/pulse/modern-ai-technologies-generative-agentic-agents-data-jha-ky8gc/

Liu, Z., Yan, S., He, W., Wan, X., & Shen, X. (2024). Beyond realities: Designing cross-reality interactions in the immersive narrative experience. In *2024 IEEE international symposium on mixed and augmented reality adjunct (ISMAR-adjunct)* (pp. 571–574). IEEE.

NTT. (2022). *Architectural concept for global trusted dataspaces: Realizing trusted data distribution across organizations and industries*. Retrieved July 25, 2025, from https://www.nttdata.com/global/en/-/media/nttdataglobal/1_files/technology/me_pr_jun_30_01.pdf

NVIDIA. (2025). *What is physical AI?* Retrieved July 29, 2025, from https://www.nvidia.com/en-sg/glossary/generative-physical-ai/

Otto, B., Ten Hompel, M., & Wrobel, S. (2022). *Designing data spaces: The ecosystem approach to competitive advantage* (p. 580). Springer Nature.

Parada, C. (2025). *Gemini Robotics brings AI into the physical world*. Retrieved October 10, 2025, from https://deepmind.google/discover/blog/gemini-robotics-brings-ai-into-the-physical-world

Qiu, R. (2024). Web3 and service science: Perfectly aligned to maximize the values of all stakeholders in the digital economy. *Digital Transformation and Society, 3*(4), 337–339.

Qiu, R. (2025). Revolutionizing digital transformation in business and society through agentic AI integration. *Digital Transformation and Society, 4*(2), 109–110.

Schuiki, L., Stach, C., Giebler, C., Hoos, E., & Mitschang, B. (2025). Enabling trusted data sharing in data spaces: PROTON-A privacy-by-design approach to data products. In *Proceedings of the 11th international conference on information systems security and privacy* (Vol. 1, pp. 95–106).

Shen, M., Tan, Z., Niyato, D., Liu, Y., Kang, J., Xiong, Z., Zhu, L., Wang, W., & Shen, X. (2024). Artificial intelligence for web 3.0: A comprehensive survey. *ACM Computing Surveys, 56*(10), 1–39.

Solmaz, G., Cirillo, F., Fürst, J., Jacobs, T., Bauer, M., Kovacs, E., Santana, J. R., & Sánchez, L. (2022). Enabling data spaces: Existing developments and challenges. *Proceedings of the 1st International Workshop on Data Economy* (pp. 42–48)

Sron, M., & Heitmayer, M. (2024). *Digital Wonderland in Web3*. Retrieved July 30, 2025, from https://digitalfashionnetwork.net/wp-content/uploads/sites/452/2024/07/DFIC1-Marta-Sron.pdf

Thaker, S., & Heitmayer, M. (2025). Exploring user perceptions of peak experience in extended reality user experiences in immersive and augmented spaces: A case study of the Jadu Web3 community. *Journal of Humanistic Psychology*. https://doi.org/10.1177/00221678251331141

Wan, S., Lin, H., Gan, W., Chen, J., & Yu, P. S. (2024). Web3: The next internet revolution. *IEEE Internet of Things Journal, 11*(21), 34811–34825.

Wu, H., Abdallah, M., Chi, Y., Lin, L., & Cai, W. (2025). Web3 multimedia applications: Under the impact of decentralization. *ACM Transactions on Multimedia Computing, Communications and Applications, 21*, 1.

Zeng, Y., Lu, E., & Sun, K. (2025). Principles on symbiosis for natural life and living artificial intelligence. *AI and Ethics, 5*(1), 81–86.

Zhang, C., Liu, Y., Xu, M., Yang, X., Li, P., Yang, C., et al. (2025). Trans-border trusted data spaces: A general framework supporting trustworthy international data circulation. *IEEE Access, 13*, 30481.

Glossary

Blockchain A foundational technology for Web3 that ensures decentralized trust and features properties like immutability and transparency.

Cloud Computing A computing model or service that enables businesses and end-users to access needed data, information, and functions over the Internet, offering advantages in infrastructure, scalability, cost-effectiveness, and accessibility.

DApps (Decentralized Applications) A foundational pillar of the Web3 ecosystem that functions as the user-facing layer (interface) and embodies Web3 principles by shifting users from passive consumers to active contributors in governance and value co-creation.

DAOs (Decentralized Autonomous Organizations) A foundational pillar of the Web3 ecosystem that revolutionizes governance with decentralized and transparent decision-making, fundamentally managed by token-based voting systems and smart contracts.

Decentralization A core principle and architectural shift where data is distributed across numerous computing nodes over a decentralized network instead of being stored on a single central server, thereby enhancing security and resilience.

DeFi (Decentralized Finance) A foundational pillar of the Web3 ecosystem that transforms traditional financial systems by promoting equitable value distribution and transparency through decentralized ledgers without relying on intermediaries.

deepservice The next generation of intelligent services that are meticulously personalized, human-centered, ethically designed, and context-aware, making the Web3 ecosystem profoundly human.

DIDs (Decentralized Identifiers) A foundational pillar of the Web3 ecosystem that redefines digital identity by offering a secure, self-sovereign, and verifiable framework, granting users control over their digital identities and reducing reliance on centralized authorities.

Immutability A core blockchain property that ensures data, once recorded on the decentralized network, cannot be altered or deleted, and guarantees the integrity, transparency, and auditability of all recorded data and transactions.

R. Qiu, *Value Co-Creation in Web 3.0*, Advances in Data Analytics, AI, and Smart Systems, https://doi.org/10.1007/978-3-032-12911-6

IoT (Internet of Things) A network framework enabling applications involving networked physical and virtual objects, enabling data exchange and communication to support intelligent identification, tracking, monitoring, and management.

Metaverse A collective, immersive virtual space, often supported by Web3 protocols and AI/ML, that is quickly gaining momentum by blurring the lines between our physical and digital lives and expanding experiential possibilities.

Privacy-Preserving Computation (PPC) Technologies and methods that underpin the Web3 ecosystem by transforming how data is collected, analyzed, and shared, while upholding security and ethical integrity (e.g., using Federated Learning and Zero-Knowledge Proofs).

Service Science A field of study whose principles, such as people-centricity and value co-creation, are integrated into the Web3 ecosystem to promote and deliver value to all stakeholders.

Trusted Data Spaces (TDSs) A foundational data and computational technology that empowers individuals and organizations to retain granular control and ownership, ensuring data is ethically shared, transparently governed, and collaboratively, efficiently, and effectively used in the Web3 era.

Value Co-creation A powerful principle in the Web3 era where value is redefined as a shared outcome of collaboration, collective governance, and continuous co-evolution among all participants, rather than predefined utility to be extracted by a centralized entity.

Web 2.0/Web2 The interactive and social web, characterized by centralized, platform-driven architectures that focus on user-generated content, interactivity, and social networking, where data is largely controlled by corporations.

Web 3.0/Web3 The decentralized, user-centric web of the future - a blockchain-based paradigm that promotes individual ownership and control over digital assets and data, creating an open, trustless, collaborative, and permissionless Internet.

Zero-Knowledge Proofs (ZKPs) A privacy-preserving technology that enables one party to prove knowledge of a fact (e.g., eligibility) without disclosing any sensitive underlying information.

MIX
Papier aus verantwortungsvollen Quellen
Paper from responsible sources
FSC® C105338

If you have any concerns about our products,
you can contact us on
ProductSafety@springernature.com

In case Publisher is established outside the EU,
the EU authorized representative is:
Springer Nature Customer Service Center GmbH
Europaplatz 3, 69115 Heidelberg, Germany

Printed by Libri Plureos GmbH
in Hamburg, Germany